AF521294

THE GAME CHANGER

Game Theory
and
Low Intensity Maritime Operations

THE GAME CHANGER

Game Theory and Low Intensity Maritime Operations

Arun Pratap Golaya

Foreword by

Admiral Devendra Kumar Joshi
PVSM, AVSM, YSM, NM, VSM (Retd)
Chairman, National Maritime Foundation, and
Former Chief of Naval Staff

National Maritime Foundation

PENTAGON PRESS

THE GAME CHANGER
Game Theory and Low Intensity Maritime Operations

Arun Pratap Golaya

First Published in 2014

Copyright © National Maritime Foundation

ISBN 978-81-8274-815-6

All rights reserved. No part of this publication may be reproduced, stored in a retrieval system, or transmitted, in any form or by any means, electronic, mechanical, photocopying, recording, or otherwise, without first obtaining written permission of the copyright owner.

Disclaimer: The views and opinions expressed in the book are the individual assertion of the Author. The book does not reflect the view of the Indian Navy or the National Maritime Foundation. The Publisher does not take any responsibility for the same in any manner whatsoever. The same shall solely be the responsibility of the Author.

Published by
PENTAGON PRESS
206, Peacock Lane, Shahpur Jat,
New Delhi-110049
Phones: 011-64706243, 26491568
Telefax: 011-26490600
email: rajan@pentagonpress.in
website: www.pentagonpress.in

Branch:
Flat No.213, Athena-2,
Clover Acropolis,
Viman Nagar,
Pune-411014
Email: pentagonpresspune@gmail.com

Printed at Aegean Offset Printers, Greater Noida, U.P.

Dedicated to
my daughters and nieces: Sanyukta, Pallavi, Neha, Ria, Nikita and Sanjana – for the wonderful children that they are and the wonderful adults they will grow up to be...

"Chess is not a game. Chess is a well-defined form of computation. You may not be able to work out the answers, but in theory there must be a solution, a right procedure in any position. Now real games are not like that at all. Real life is not like that. Real life consists of bluffing, of little tactics of deception, of asking yourself what is the other man going to think I meant to do. And that is what games are about in my theory."

—John von Neumann

Contents

Foreword

A logical analysis of all the available options to identify the most favourable choice is an intrinsic part of the human decision making process. The advancements in mathematical modelling help solve problems and provide the framework for complex situations where fixed and variable factors exist in equal measures. A logical reasoning mechanism also helps to showcase the underlying thought process and the rationale adopted for reasoning.

Amongst the various decisions-making methodologies, Game Theory is unique and offers a holistic view of mutually dependent and interrelated factors. The theory is premised on the existence of a strong element of interdependence among the decisions taken by competing entities. The basic assumptions of rationality and mutual interdependence are the two key characteristics of players, who may be individuals, groups of individuals or even nation states.

From the security prism, the initial applications of Game Theory were limited to solving the larger strategic problems. This was especially true in the interactions between the nuclear powers, where Game Theory was routinely used to explain deterrence concepts. Application of Game Theory for combating terrorism and piracy is a more recent development, particularly in the aftermath of 9/11 attacks in the United States. The study of Game Theory has grown substantially and academic examination of the problem of terrorism from a Game Theory perspective is fairly robust. The exchange between terrorists and the security forces cannot be analyzed as though one side is passive. In essence, the decisions taken by the security forces are closely linked with those of the terrorists and this is a continually evolving and mutating process. This makes Game Theory an ideal tool for studying the problem and formulating possible solutions to complex and diffuse problems.

Information asymmetry is another important facet of Game Theory and is relevant to interactions between security agencies and terrorists/non-state actors. 'Actionable intelligence' is hard to come by, especially in the maritime domain, thus impacting on the efficacy of the probable responses. In such instances, Game Theory offers a 'predictive solution' by identifying the possible (and feasible) options by using appropriate tools. Game Theory also has much to offer in countering piracy and other Low Intensity Maritime Operations. A wider understanding of Game Theory amongst the policy makers, planners, as well as practitioners of security can help in making our seas and our coasts safer.

I am aware that Game Theory, as a subject, is taught to the military officers during their training, but at a relatively senior level. This leads to a large number of personnel remaining unaware of the basic conceptual tenets. Further, most of the existing academic work assumes *a priori* knowledge of these concepts amongst the readers and the uninitiated is often left struggling.

This book emerges from the second Admiral AK Chatterji Fellowship, constituted to improve the 'maritime consciousness' among the analytical and scholarly communities. It is an excellent attempt to fill this void by explaining the basic concepts of Game Theory and link them to actual applications in the maritime domain. It is hoped that the book would initiate the 'generic security community' to the world of Game Theory especially amongst the younger strategists and naval officers.

Admiral Devendra Kumar Joshi
PVSM, AVSM, YSM, NM, VSM (Retd)
Chairman, National Maritime Foundation, and
Former Chief of Naval Staff

Preface

'Strategic thinking is the art of outdoing an adversary knowing that the adversary is trying to outdo you.'

—Dixit & Nalebuff (1991)

Terrorism, piracy and Low Intensity Maritime Operations are certainly not games...and neither are the responses to thwart them. The people involved, on the 'good' as well as the 'bad' side, are nonetheless locked in an endless mind game with each trying to out-think the other.

'Game Theory' is a branch of applied mathematics that is concerned with how rational individuals make decisions when they are mutually interdependent. Game Theory has been used extensively in fields ranging from economics and political science to psychology and biology. The *'game'* involves two or more players making decisions so as to maximise their returns. In attempting to do so, they must anticipate the response of their adversaries, sometimes even in the absence of precise information. It is essentially about *'outguessing'* the adversary, by putting yourself in the opponents' position. Since a logical adversary would presumably do the same—the interplay between the players and the choices they must make often mirror real-life situations.

This structured study of how people with opposing interests interact in a conflict scenario can provide a framework to study not only the larger strategy issues but also low intensity conflicts and for defence against terrorist attacks. Scholars worldwide have written extensively on the subject and many have advocated Game Theory as an ideal tool for combating terrorism. Sandler, Tschirhart, and Cauley had put forward a 'Game-Theory' centric model of negotiation with terrorists as far back as 1983. The same approach has since been used by many others and application of Game Theory to the study of terrorism has grown significantly following 9/11.

The primary purpose of this book is to introduce the *concept* of Game Theory to the uninitiated reader, with the focus on application of Game Theory to combat terrorism, piracy and for Low Intensity Maritime Operations. Whilst the subject itself involves a fair amount of mathematics, the attempt here has been to emphasise the underlying logic and strategies rather than the full mathematical details.

I can already see the purists cringe at the idea of a book about Game Theory without mathematics and say that it would be like a book about botany with no plants! My only response other than raising both my hands and saying *mea culpa* would be...I am not trying to write a book about botany; Rather I'm trying to show people that plants can be interesting...and useful. Readers, who find the subject intriguing, can pick up a good botany book from the many available!

Similarly a number of authoritative writings on the application of Game Theory to the field of combating terrorism and piracy are also available. The problem frequently is that the papers have been written by subject matter experts who assume a basic knowledge of the concepts amongst the readers. This may always not be the case. The book will therefore, after introducing the basic concepts in Section I, review some of the existing studies on the subject in the second section. Once again, here an attempt has been made to explain the concepts and practical applications without going into the mathematical details. The emphasis therefore is on highlighting the format of the 'games' and the conclusions reached by the studies. The original studies must be referred to for the exact mathematical details. This section also contains some stand-alone essays primarily adapting some of the existing studies to the Indian scenario in attempt to glean policy recommendations.

Acknowledgements

At the outset I would like to thank the Admiral AK Chatterji Fellowship under the National Maritime Foundation (NMF) for giving me the opportunity to undertake this work. The NMF must also be thanked for supporting me at all stages of my progress. What started off as a primer on Game Theory, slowly transformed into this monograph as elements of how the concepts could actually be used were incorporated. The constant encouragement of two successive Directors of NMF, Vice Admiral PK Kaushiva (Retd) and Dr Vijay Sakhuja must specifically be acknowledged.

My colleagues at the NMF, while I was posted there, including Commander KK Agnihotri and Commander Raghvendra Mishra must also be thanked for helping me with their vast experience as researchers. We spent many afternoons discussing not only this work but a large number of issues concerning matters maritime. I have always enjoyed picking their brains and, in the process, learning from them.

I must also acknowledge the many serving and retired senior naval officers who were supportive and gave me their valuable time to discuss the issues. I would like to particularly mention Rear Admiral Raja Menon (Retd), Commodore C Uday Bhaskar (Retd) and Rear Admiral SY Shrikhande—whose guidance and encouragement I must gratefully acknowledge. I also drew from the vast experience of a number of individuals, too numerous for me to completely list here due to the constraints of space. All of them, to varying degrees, helped me refine my work.

My wife Durgesh and daughters Sanyukta and Pallavi were a constant source of encouragement and support.

Last, but certainly not the least, I record my gratitude to my parents and siblings who during my formative years inculcated in me a love for reading. Even as I often ignored school work, preferring rather to spend time with books or trying out childish science experiments, they decided to encourage rather than curb my 'originality' as *papa* often put it.

SECTION ONE
GAME THEORY

CHAPTER 1

Introduction to Game Theory

Game Theory has been defined as "the study of mathematical models of conflict and cooperation between intelligent rational decision-makers".[1] The *Compact Oxford English Dictionary* defines Game Theory as "The mathematical study of strategies for dealing with competitive situations where the outcome of a participant's choice of action depends critically on the actions of other participants". Game Theory finds application in many diverse fields and has been keenly researched. Eight Game-Theorists have won the Nobel Memorial Prize in Economic Sciences. Robert Aumann, the inaugural president of the Game Theory Society in his presidential address stated that 'Game Theory' is an unfortunate term for a study that might be better understood as 'interactive decision theory'.

Logical analysis of available options as an approach to 'problem solving' has been around since time immemorial. The basic premise of Game Theory, namely that of interdependence of decision making in an adversarial (or at least competitive) scenario is applicable to a number of real life situations. Mathematical models to help solve the problems can be useful but are not absolutely critical as long as the underlying logic is correct. Though many scholars before him had used the concepts of game thoery to put across their views, Émile Borel in his book *Applications aux Jeux de Hasard* had for the first time introduced the scientific and mathematical study on the subject. Modern Game Theory began with the Hungarian-American mathematician John von Neumann providing proof of 'equilibrium' in 'two-person', 'zero-sum' games (description of basic Game Theory terms is given in the next few pages). This theory was developed extensively in the later years by many scholars and found application in a variety of fields.

Concepts of Game Theory have also found way into popular culture and have been used (knowingly or unknowingly) in many popular books, movies and television shows. In the classic 'The Hunt for Red October', for example, Captain Ramius informs his officers that he has sent a letter to the Soviet command, informing them of his plan to defect. In Game Theory terms such a step would be called 'burning bridges'. Burning a bridge would leave the player with no option of turning back on a decision that has already been taken. The same movie also talks about "The hard part about playing chicken" being "knowing when to flinch". Chicken, in fact, is a well known game (as well as a concept of Game Theory) dealing with 'who will blink first'. It was supposedly extensively used during the Cold War era to work out the possibilities (and acceptable thresholds) in eyeball-to-eyeball confrontations.

Game Characteristics

The primary difference between Game Theory and other decisions-making methodologies is that Game Theory does not see a decision in isolation. It is concerned therefore with how rational, and interdependent, individuals (or groups of individuals) make decisions. 'Rationality' and 'interdependence' are thus the two key characteristics of players that form the basis of all games.

Rationality implies that individuals may be assumed to act in their own self-interest. The same is a logical assumption but may not always be completely true. The problem frequently is that some real-life situations may be so complex that individuals may be taking 'wrong' decisions, unfavourable to their interests, simply because they have inadequately analysed the complete scenario. Rationality, whilst assuming that individuals would prefer actions that are favourable to their interests, presupposes that individuals can always determine, probabilistically if not exactly, the outcome of all their actions. This seems unrealistic if we consider the complexity of most decisions, and the amount of information that often needs to be analysed. Solving complex decisions in real life often involves adopting simplistic rules that are generally suboptimal. Prejudices, wrong assumptions, incomplete information or the inability to process the complex information often make decisions seem 'illogical', at least in hindsight. Game Theory generally assumes that individuals act as if they were fully rational. It is therefore not concerned with how individuals *actually* solve complex decision making problems but with how they would *optimally* solve the problems.

Mutual interdependence implies that Game Theory is concerned with situations where individuals are dependent, at least partially, on the decisions made by others. In this situation the benefits that accrue to any one individual are determined not only by his rational decision making but equally by the actions of other players in the game. Whilst making decisions, the interdependent individuals must also know how to anticipate the effect of their own actions on the behaviour of others. Each individual determines the optimal decision in order to achieve the most desirable outcome based on how other mutually interdependent individuals behave.

'*Games*' or interactive decision making situations would normally have the following characteristics:[2]

Conflict: To begin with there should be a conflict or a bone of contention. In any game, the participants (who may be individuals, groups, countries etc) who are known as 'players' must have interests that are contrary to each other's. These players, whose interests are at cross-purposes, therefore try to maximise their own gains, or 'payoffs', in the given situation. Any person or entity who makes a choice in a game or who receives a payoff from the outcome of those choices is a player.

Adversarial Reaction and Interaction: Each player can take some decisions and thereby has some control over the outcome of the conflict. The decision one player takes can impact the course of the game and further affect the options available for the adversary.

Outside Forces: Some aspects of the conflict are outside of the players' control and may be governed by chance. Some aspects can similarly be unknown to the players.

These characteristics apply to quite a few real-life situations and the solutions can thus be worked out by use of Game Theory. For our purpose, these characteristics quite obviously apply to anti-terrorism,[3] anti-piracy and the like. William Poundstone called Game Theory "a study of conflict between thoughtful and potentially deceitful opponents". This definition aptly applies to terrorists as well as to the state, which may be considered deceitful by the non-state actors.

Basics of Games Theory

To get a basic understanding of the concepts, let us first familiarise with some of the terms commonly used in Game Theory. Whilst describing the terms, parallels have been drawn from other fields including common games such as chess, cricket or card games for the ease of understanding. This over-simplification, whilst not being absolutely accurate under all circumstances, has been used throughout this book to explain the basic concepts.

Strategy: In a game each player chooses from a set of possible actions, known as strategies. Strictly speaking, the strategy used by any player in a game is the complete 'action plan' used by the player for dealing with different situations that may arise. The strategy adopted will determine the player's behaviour and will further determine the action that the player will take in a given situation. The term is often confused with a 'move'. A move is an action taken by a player at some point during the game. For example, in a game of chess, moving the Bishop from a particular position to another would be a move. Strategy, on a broader level, is the complete algorithm for playing the game which tells a player what to do for every possible situation faced at different stages of the game. Attempting to maintain 'status quo' in any situation or a policy of 'non negotiation' with terrorists would be examples of strategies. 'Tit for tat' would be another example of a strategy in Game Theory where a player initially cooperates and subsequently responds by replicating the opponent's previous action. If the opponent was cooperative in the last move the player using this strategy would also cooperate. A 'pure strategy' provides a complete definition of how a player will play a game (the move a player will make for any situation faced). The set of pure strategies available to a player is called that player's 'strategy set'. A 'mixed strategy' is an assignment of a probability to each pure strategy. This allows for a player to select a pure strategy from amongst many available, thus giving an infinite number of mixed strategies even if the strategy set is finite.

Common Knowledge: Knowledge that all the players have about the structure of the game itself is termed common knowledge. In formulating the games, it is assumed that not only does each player know the rules (and the structure) of the game but also that the player knows that the opponent knows them too. Further the player knows that the opponent knows that he know the rules… and so on *ad infinitum*. The term 'common knowledge' avoids spelling out the infinite recursion to which this logic may be extended. Information is said to be common knowledge if it is known to all the players, if each player knows that all the other players know it, further if each player knows that all the players know that all the other players know it, and so forth *ad infinitum*. Again to take an example from chess, a game would be impossible unless the participants knew not only how a bishop is permitted to move, but also knew that the opponent knows it as well. A person who lacks common knowledge about the format of the game cannot obviously be a player.

Symmetric Game: A game is said to be symmetric if the payoffs for playing a particular strategy depend only on the strategies employed by the opponents and not on who is employing them. To put it simply, if one can change the identities of the players without changing the payoff to the strategies, then a game is symmetric. A symmetric game is 'fair' to all players as all players would have an equal chance of 'winning'. Games between state and non-state actors would, in most cases, not be symmetrical.

Zero-sum Game: A game where any participant's gain is exactly matched by the losses of the other participant(s) is called a zero-sum game. This is a situation in which one player can gain (or lose) only at the expense of the others. It is so called because when the total gains of all the participants are added up, and the total losses subtracted from this figure, they will sum to zero. A zero-sum game is also called a *strictly competitive* game. Thus sharing a cake, where taking a larger piece reduces the amount of cake available for others, is a zero-sum game if all participants desire each unit of cake equally. It ceases to be a zero sum game if one of the participants prefers taking the piece with a cherry on top even if it is smaller.

There can be no win-win situations or lose-lose situations in zero sum games. In contrast, non-zero-sum describes a situation in which the sum total of the gains and losses are either less than or more than zero. To take an example from cricket, if winning a match gives a team two points, losing zero and the two teams share a point each in case of a draw (or in the event of a match being called off due to rain)—each individual match will be 'zero-sum' or strictly competitive. To take the example further, in the World Cup, a particular team may have already accumulated sufficient points at the league stage to enable an entry to the finals with a simple draw whilst the other may be in a do-or-die situation needing a win to qualify. In the overall scheme of things, the former team may derive a greater value from a draw—and the last match therefore ceases to be zero-sum.

Simultaneous and Sequential Games: All participants in a 'simultaneous game' take decisions simultaneously with each player choosing an option without any knowledge of the option(s) being chosen by others. It is important to note that simultaneous refers not to the *time sequence* but the knowledge the opponents have of a player's moves. A game where the moves are not chronologically simultaneous, but where the later players are not aware of the earlier players' actions, may also be considered *effectively* simultaneous. A sequential game, on the other hand, is a game where one player chooses his action before the others choose theirs. The later players must have some information of the preceding player's choice; otherwise the difference in time would have no meaningful effect. The choice of the second player is obviously based on the decision made by the first, and would likely to be influenced by it, more than the other influencing factors. A game of chess, once again, is an example of a sequential game where each move is based on and dependent upon the previous move by the opponent. Another example of a simultaneous game could be the tendering process wherein the bidding firms are asked to submit quotes in sealed envelopes. Since they do not know the bids of the competitors, the game is effectively simultaneous even though the bids could have been submitted at different times. In contrast, the auctioning process where the players may openly bid and these bids are not only observed but also reacted upon by the other players is sequential. Simultaneous and sequential games are also sometimes referred to as 'static' and 'dynamic' games respectively.

Repeated Games: A repeated game, also sometimes called a super-game or an iterated game, is a game which consists of a number of repetitions or iterations of some base game (called a stage game).

Cooperative and Non Cooperative Games: A cooperative game is a game where groups of players may form 'coalitions' or teams, wherein cooperative behaviour can be enforced. Hence the game is a competition between *coalitions* of players, rather than between individual players. It is important to note that a game is classified as *cooperative only* if the players' commitments are binding on them and the system requires them to adhere to their promises. Games are rarely cooperative because of the difficulty in enforcing coordinated behaviour. Such mechanisms, however, are common in real life situations, for example in legally binding contracts. It is often assumed, though incorrectly,[4] that *communication* among players is allowed in cooperative games, but not in non-cooperative ones. Some games may also have characteristics of both cooperative and non-cooperative games, for example, coalitions of players may be formed as in a cooperative game, but the players may play in a non-cooperative fashion. Such games are usually termed '*Hybrid*' games. Non-cooperative Game Theory does not prevent individuals from cooperating or working together, it merely assumes that this will only happen if individuals perceive such cooperation to be in their individual self-interest. In other words they cooperate because they *want to* and not because they *have to*. Rationally, they would want to cooperate only if it furthers their individual gains. Since in non-cooperative Game Theory the players are unable to enter into binding contracts with one another, this branch of Game Theory is inherently individualistic.

Perfect Information: Players make decisions based on information available to them. This 'information' could be accurately known but would more often be inaccurate, sketchy, a guess or a mathematical probability. The term 'perfect information' describes a situation when a player has available the same information to determine all the possible moves as would be available at the end of the game. A game of chess, where each player can see all the pieces on the board at all stages of the game, is an example of a game being played under conditions of perfect information. Follow-the-Leader is another example of a game of perfect information. Most card games, where each player's cards are hidden from others till they are played, are examples of games of imperfect information.[5] A government trying to prevent a terrorist attack when the actual target or the date of the attack is not known similarly lacks perfect information. Perfect information must be differentiated from a similar concept called 'complete information'. Complete information requires that every player knows the strategies and payoffs available to the other players but this does not necessarily mean that knowledge of the actions taken is also available. The adage 'hindsight is an exact science' applies well to most imperfect games (and real life situations) where decisions seem illogical when an event is analysed *post facto* but were quite logical when only incomplete or imperfect information was available. Information that is available to one of the players may not be known to the others. A situation where no player has information different from that available to other players at the time of making a move, or at the end of the game, is also sometimes called 'symmetric' information. Obviously, any game of incomplete or asymmetric information is also a game of imperfect information. Some outcomes or decisions may also depend upon factors which are beyond the control of either of the players. Again to take an example from cricket, whether it rains during a match or not is certainly beyond the control of the players. If it does rain, the team batting second, for example, may be affected differently than the team batting first. The occurrence of rain, and the specific time during the match that it actually does, may be attributed to an external player 'nature'. This player, even though not having any stakes in the final outcome, can influence the course of the match through the 'moves' made by it. If this player makes no move or if the moves made by nature can be equally observed by every other player in a game and they are equally affected, such that the element of 'chance' has little role in the outcome, the game is said to be of 'certain' information. A game of certainty can be a game of perfect information if it has no simultaneous moves.

Strategic thinking involves not only predicting what the other player will do, but equally guessing the information available with the adversary, based on which the decisions will be taken. If the moves are simultaneous, this has little relevance. In any sequential game however, information and its categories such as certain, perfect or complete become critical. By observing the moves of the adversary in a sequential game, a player acquires some of the information (based on how the adversary acted) before taking a decision about the next move.

Representation of Games

Games are normally represented mathematically using methods which give a gist of the set of players, a set of moves available to those players, and the resulting payoffs for each combination of moves. Most cooperative games are presented in the function form, while the 'extensive' and the 'normal' forms are used to define non-cooperative games.

The *normal form* (also called the strategic form) represents the game by using a matrix. The players are assumed to act simultaneously (or alternately, without knowing the actions of the opponent when acting sequentially). Let us assume a two-player, two-strategy game. The two players have two options available to each. Player 1 has to select either Option A or Option B and Player 2 has Option C or Option D as the available choices.. The player 1 gets a payoff of say '2' if he takes Option A and his opponent chooses Option C. The opponent's payoff in this situation is, say, '3'. The same as well as other combinations can be represented as a matrix (figure 1.1).

	PLAYER 2 CHOOSES OPTION C	PLAYER 2 CHOOSES OPTION D
PLAYER 1 CHOOSES OPTION A	2,3	1,4
PLAYER 1 CHOOSES OPTION B	3,2	4,1

Figure 1.1: Normal form matrix for a 2-player, 2-strategy game

The following conventions are followed in matrices worked out using the normal form:

- When there are two players; one is represented in the rows and the other in the columns.
- The strategies available to each player are specified by the number of rows and the number of columns (two rows for the two available strategies in the example above).
- The payoffs are provided inside the matrix. The first number is the payoff received by the row player (Player 1 in our example); the second is the payoff for the column player (Player 2 in our example).

All games can obviously not be represented as a simple matrix. Especially in the case of sequential games, the exact stage of the game when a particular decision is taken becomes critical. In sequential games greater attention therefore needs to be placed on the timing of the decisions to be made, as well as on the amount of information available to each player when that decision needs to be taken. This type of game is usually represented not with a matrix but with a decision (or game) tree, with each decision point being termed a node. This representation of a game is called the *extensive form.* The extensive form, as the name suggests, allows a far more explicit representation of the game. If a game is complex, being played sequentially or when the players have some information about the choices made by their opponents, the game is usually presented in extensive form. This form can represent a number of important aspects, like the sequencing of players' possible moves and the choices available to them at every decision point. These aspects of the game are represented as decision trees (figure 1.2). In the tree, each vertex (or node) represents a point of choice for a player. The player is specified by a number listed by the vertex (in the figure 'Player 1' has been written, normally only the number '1' would suffice). The lines or branches out of the vertex represent the possible options for that player. The payoffs, for each individual chain of possible outcomes, are specified at the bottom of the tree. For example Player 1, starting first, has been assumed to have two options, namely to 'act' or be passive ('do not act'). Player 2 sees Player 1's move and again has two options. Once again they have been taken as 'act' or 'do not act', though the choices available at each node need not be the same for the players or even for a particular player. The options for Player 2 have been abbreviated to 'A' and 'D' respectively. Suppose that Player 1 chooses A and then Player 2 also chooses A, then Player 1 gets a payoff of 3 and Player 2 gets 2. If Player 1 had chosen A, but Player 2 chose D, the payoffs for both the players would be zero. The extensive form can also capture simultaneous-move games and games with imperfect information.

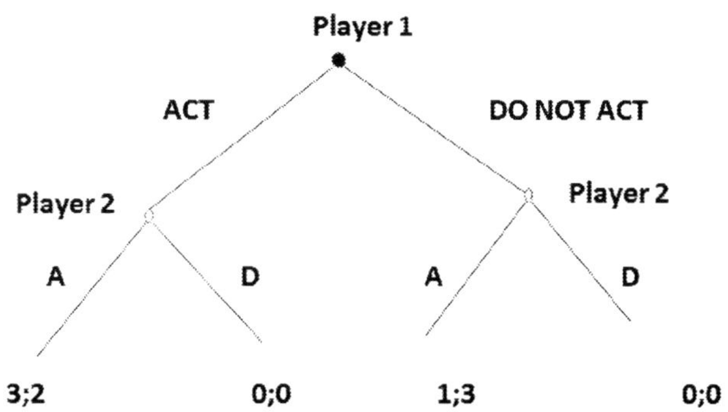

Figure 1.2: Extensive form game

A 'node' is a point in the game at which any of the players (or the player 'Nature') takes an action or a decision. The point where the game ends is also a node. Each node may have predecessor and a successor node. A predecessor to any node (say node X) is a node that must be reached before that particular node, X, can be reached. Similarly, a successor to a node X is any node that may occur later in the game if that particular node X has been previously been crossed. A node with no predecessors is termed a starting node and a node with no successors is called an end node (or end point). The decisions or actions that emanate from a node are termed branches. The sequence of nodes and branches leading from the starting node to an end node is called a path. Obviously multiple paths are available in any game due to the multiplicity of actions that may be taken and other variables. The extensive form is therefore a configuration of nodes and branches running without any closed loops from a single starting node to its end nodes (which would be more than one). By convention, an indication is provided of which node belongs to which player.

Strictly speaking, the game tree is not the same as the extensive form. The term 'game tree' is a more generic term than 'extensive form'. In the extensive form, the payoffs for each player are described at each end node whereas in a game tree the same may be replaced with the outcomes at each end node. If the final outcome is defined as the payoff profile, with one payoff for each of the players, then the two terms become synonymous.

Some games may also be described in the form of timelines. Chronologically depicting the order of events, is particularly useful for games with continuously evolving strategies, and changing information. It is important to note that the chronological sequence of the time line depicts the order of events and not necessarily the passage of time. Decision time and the interval over which physical actions are taken (real time) therefore need to be differentiated.

Dominance, Equilibriums and other Solutions

Strategic Dominance, or simply dominance, is a term used to describe a situation when one strategy is better than another strategy for any player, no matter how the opponent(s) may play. In the example in figure 1.1, player 1 would gain more on choosing option B regardless of the choice made by player 2. Many simple games can be solved using dominance.

Nash equilibrium, named after John Forbes Nash, is a solution concept in which no player has anything to gain by changing his/her own strategy unilaterally. If each player has chosen a strategy (and each player is assumed to know the equilibrium strategies of the other players), and no player can benefit by changing his or her strategy while the other players keep theirs unchanged, then the current choices represent a Nash equilibrium. Simply put, players are in equilibrium if a change in strategy by any one of them would lead that player to earn less as compared to retaining the current strategy.

Pareto optimality, named after Vilfredo Pareto, is a measure of efficiency in the solution obtained. An outcome of a game is Pareto optimal if there is no other outcome that makes every player at least as well off and at least one player strictly better off. That is, a Pareto optimal outcome cannot be improved upon without hurting at least one player.

Hicks optimality, named after John Hicks, is a measure of total efficiency. An outcome of a game is said to be Hicks optimal if there is no other outcome that results in greater total payoffs for the players. Thus, a Hicks optimal outcome is the point at which the sum total of payoffs for all the players is maximised. Obviously, a Hicks optimal outcome will also necessarily be Pareto optimal.

More often than not—a Nash equilibrium is not Pareto optimal—implying that the players' payoffs can all be individually increased. It is therefore important to note that Nash equilibrium does not necessarily represent the highest payoff for a player. More than one equilibrium in a game is also sometimes possible. These concepts will become clearer with a study of some of the common games, which have been covered in the next chapter.

NOTES

1. Myerson, Roger B., *Game Theory: Analysis of Conflict* (Cambridge, Mass: Harvard University Press, 1991), p. 1.
2. Williams, J.D., *The Compleat Strategyst: Being a Primer on the Theory of Games of Strategy* (New York: McGraw-Hill, 1966).
3. Ronald D. Fricker, Jr. *Game Theory in an Age of Terrorism: How Can Statisticians Contribute?* Department of Operations Research, Naval Postgraduate School, Monterey, CA, United States.
4. Harsanyi, John C., "An Equilibrium-Point Interpretation of Stable Sets and a Proposed Alternative Definition." *Management Science*. 20.11 (1974): 1472-1495.
5. Thomas, L.C., *Games, Theory and Applications* (Mineola, N.Y.: Dover Publications, 2003). p. 19.

CHAPTER 2

Common Games

Some common games used (and often quoted) in Game Theory literature are briefly covered in this chapter. One can often find references to these games in studies employing Game Theory. Since these games are well known and have been extensively discussed, most discussions would consider a statement such as 'a variation of the *Prisoner's Dilemma*' to be self explanatory. It is therefore important for a beginner to understand these games, which would also help in clarifying some of the concepts discussed earlier.

Before going into the common theoretical games and their variations, let us see one example of a 'game' from popular culture which has been deemed (as per one reviewer), the most outstanding example of Game Theory.[1] The situation in the climax from the Clint Eastwood film 'The Good, the Bad and the Ugly' involves three men—each with a gun, and each an excellent shot. The *players* Tuco (Ellie Walach), Angel/Sentenza (Lee Van Cliff) and Blondie (Clint Eastwood) are competing for high stakes—a *payoff* of a treasure or certain death. The men are shown standing in a triangle with a rock at the centre of the three. The money is hidden in a certain tomb, whose name Blondie had supposedly written on that rock. The players are equal, in the sense that each man can shoot another, and none can possibly fire quickly enough and kill the two others before being shot at himself. It is up to each man to work out his situation, and as the camera pans slowly moves across their faces, the audience is left spellbound wondering what each would do. *Truels* (neologism for a duel between three opponents) such as this have been studied using Game Theory and the mathematical reasoning for the-logical-thing-to-do in variations of the game have also been extensively studied. Believe it or not, but the logical thing might be (in some cases[2]) to deliberately miss! Let us examine this particular variation from the logical viewpoint:

- If the other two cannot trust Blondie to have actually written the correct location, they would realise that they cannot shoot him without risking loosing the treasure.
- This implies that the other two can only shoot each other, but only one will probably hit before the other. One opponent would still be left for Blondie.
- The other two can at the most reduce the claimants to the treasure by one. Keeping Blondie alive is critical to their own self interest.
- Blondie also, whilst being certain that he won't be shot at initially, doesn't know which of them to shoot at (best option is to shoot the person who survives between the other two).

Of course in the movie, the game was decided before it started. In a brilliant example of players making decisions based on the information available (*imperfect information*)...and other players manipulating the same, Blondie had given one adversary an unloaded gun. He had thus setup a situation (with *complete information*) where each evaluates their possible moves, but in reality, he had already won the game as he manipulated them so that the opponent with the loaded gun shot at the one with the unloaded gun whilst he himself shot the one with the loaded gun, secure in the knowledge that neither would shoot at him.

Prisoners' Dilemma

The classic game, *The Prisoners' Dilemma*, probably the best known amongst all games, involves two prisoners being interrogated separately. It was proposed around 1950 by Merrill Flood and Melvin Dresher.[3] It is assumed that the two prisoners cannot communicate with each other. The police have insufficient evidence and try and work a deal with the two. If one implicates the other, he will go free while the other receives the maximum penalty. If both remain silent, they can be charged and convicted, at best, with a lesser crime since there is insufficient evidence with the police. Yet, if both confess, none can go scot free; however neither will get the maximum penalty. What should each prisoner do?

At first sight, the answer is obvious—both should stay silent. Let us call this strategy cooperate (or 'C'). The problem is that each does not know what the other would do. We assume that each values his personal freedom more than the well-being of his partner in crime; and may thus be forced to think—"what if I remain silent but he confesses?". Each is then tempted to cooperate with the law. Let us call this strategy defect (or 'D'). To work this out in a matrix we will assume the maximum penalty to be seven years and the penalty for the alternate lesser crime that they can be charged with to be two years. If both confess, each will get five years. The payoff for each situation is taken as the time one gets lesser than the maximum penalty. The payoff for the full term is then 0, for two years imprisonment is 5 and for no jail time is 7. The matrix (figure 2.1) then becomes:

	PRISONER 2 CHOOSES OPTION C	PRISONER 2 CHOOSES OPTION D
PRISONER 1 CHOOSES OPTION C	5,5	0,7
PRISONER 1 CHOOSES OPTION D	7,0	2,2

Figure 2.1: Prisoners' Dilemma

It is of course assumed that both men are 'rational' and both men want to minimise the time they spend in jail. Not confessing (or cooperating with each other) gives them both together the best result or highest payoff for the group (5 + 5 or 10).

The 'rational' Player 1 may, however, reason as follows:

- Player 2 can confess or he can keep quiet.
- I can't know for certain that he won't confess.
- Suppose he confesses. Then I get 7 years if I don't confess, 5 years if I do, so it's better to confess.
- On the other hand, if he doesn't confess, but I do—I can go free. If I also don't confess, I still get 2 years. So it's again better for me to confess.
- Either way, regardless of what he does, it's best for me if I confess.
- Therefore, I'll confess.

Player 2, also being rational, can and probably will reason in much the same way—in effect making them both confess and go to prison for 5 years each. Yet, if they had acted 'irrationally', and cooperated with each other rather than with the police by keeping quiet, they could have got off with 2 years each.

This game illustrates a conflict between individual and group rationality. A group whose members individually pursue rational self-interest may all end up worse off than a group whose members act contrary to rational self-interest. This may be explained by concept of 'dominant strategy equilibrium'. Since to confess is the dominant strategy for both the players—they are both in a dominant strategy equilibrium. To confess is the dominant strategy for each, and when both prisoners confess, it leads to a dominant strategy equilibrium. This is also a *Nash Equilibrium* for this game because given that the other player has confessed, each individual regards his own choice (confessing) as optimal. *No player has an incentive to independently change his own strategy.* It is a dilemma because each prisoner would have been better off if neither had confessed. Independent rational choices do not always achieve Pareto optimal results.

This remarkable result—that individually rational action results in both players losing out—has had wide application in human interaction. There are many interactions in the real world that seem very similar, from arms races to traffic congestion and depletion of resources by over-exploitation. Whilst these are all quite different, the underlying logic—for individually rational action leading to inferior results for each party—remains valid.

On the other hand, many detractors feel that Prisoners' Dilemma is theoretical, over-simplified and a possibly unrealistic concept. Compelling as the rationale behind the dominant strategy equilibrium is, it is not the only way this problem might be reasoned out. It may also not really be the most rational answer after all. A number of issues criticising the game or proposing alternative variations can be raised. This has led to a large volume of scholarly literature on the subject. The following needs to be highlighted:

- The same underlying principle can equally well apply to multiple player interactions.
- It has been assumed that no communication is permitted between the two prisoners. If they could communicate and coordinate strategies, the results would be different.
- Similarly even if they are not permitted to communicate but if each knows the other person well (due to having worked together before) so as to have developed a level of mutual trust, we would again obtain a different outcome.
- In the Prisoners' Dilemma, the two prisoners interact only once. Repetition of the interactions would give interesting results.

Many variations of the game have been developed. One of these, called deadlock, is similar to the Prisoner's Dilemma except that while each player hopes for the other to cooperate, neither is prepared to do so. Unlike the basic Prisoner's Dilemma, the resulting equilibrium in this case would be Pareto optimal.

Stag Hunt

The *Stag Hunt* based on a situation propounded by the Genevan philosopher Jean Jacques Rousseau (before modern Game Theory evolved) is another interesting exercise in social interaction. The game involves two hunters who can either individually hunt a smaller animal (a rabbit) or collaborate to catch a bigger animal (a stag). Hunting a stag would require cooperation, with minimal chances of success for each individual hunter acting in isolation. The rabbit can be captured regardless of what the other person does. The 'stag hunt' is therefore the most beneficial option for the group but requires them to collaborate. Before the hunt starts, the players can discuss strategy but no player can be certain that the companion will abide by the agreement.

Let us again assume two players with two possible strategies—cooperate ('C') or hunt alone ('H'). The individual payoff for a rabbit is assumed to be 5 and for the much bigger stag 10. The payoff for a player trying to hunt the stag would be zero if his companion decided to hunt a rabbit instead. Hunting a stag is therefore more risky, even if it has greater benefits.

It makes sense for either player to pick a stag if his companion chooses a stag, and to pick a rabbit if the companion chooses a rabbit. It stands to reason that either of the two options is equally likely. Both hunters would either jointly go for the stag or both would go for the individual rabbits. This game thus leads to two possible pure strategy equilibriums (figure 2.2). Both players should logically prefer the higher equilibrium (10,10) to the other one as it is both Pareto optimal and Hicks optimal. However, the lower equilibrium (5,5) may be less risky as the payoff dependence on the other player's strategies is lower. Therefore, the first equilibrium is *payoff-dominant* while the other is *risk-dominant*.

	HUNTER 2 CHOOSES OPTION C	HUNTER 2 CHOOSES OPTION H
HUNTER 1 CHOOSES OPTION C	10,10	0,5
HUNTER 1 CHOOSES OPTION H	5,0	5,5

Figure 2.2: Stag Hunt

An interesting variation of the game involves a pack of wolfs. Some wolfs amongst the pack can also 'do nothing' except joining in at mealtime. Even assuming that the others are willing to share, the animal hunted with only some group members contributing would be smaller, thereby reducing the share per participant or free-loader. This is a good solution from the social viewpoint as it demonstrates that even selfish incentives can produce social cooperation. The game becomes even more interesting if the effort taken to cooperate is subtracted from the payoffs and 'free loading' may therefore lead to some returns which are higher than the returns per individual from the larger prey. Variations of the game find extensive application in international relations.

Chicken

The game chicken has often (some feel incorrectly) been applied to international eyeball-to-eyeball scenarios. The game in its simplest forms involves two drivers (teenagers with something to prove?) speeding towards each other down a narrow road. The first to swerve loses face. If neither swerves, however, they can both die. The basic principle of the game is that while each player prefers not to yield, the worst possible outcome occurs when both players do not yield.

	DRIVER 2 STAYS	DRIVER 2 SWERVES
DRIVER 1 STAYS	-100, -100	1,-1
DRIVER 1 SWERVES	-1,1	0,0

Figure 2.3: Chicken

The game again leads to two equilibriums (figure 2.3) as was the case with the stag hunt. The question as to which is more probable (even plausible) is also something that is common to both games. But that is where the similarity ends. In stag hunt even non-cooperation gave a favourable (though lesser) payoff, this is not the case with chicken.

Many analysts have compared the Cuban missile crisis of the Cold War era and the Indo-Pak confrontation during Operation Parakram[4] to the classic game of chicken. Bertrand Russell famously compared the game of chicken to nuclear brinkmanship when he said "As played by irresponsible boys, this game is considered decadent and immoral, though only the lives of the players are risked. But when the game is played by eminent statesmen, who risk not only their own lives but those of many hundreds of millions of human beings...".

Chicken, with the result 'no one wins' unless both compromise, is the usual game used to model conflicts in which the players are on a collision course. The concept of Mutually Assured Destruction has also therefore been compared to this game. It is interesting that Theodore Sorensen, special counsel to the US President used the language of 'moves' to describe the deliberations by key advisors to President Kennedy during the Cuban missile crisis.

Whilst international confrontations can certainly be studied from the Game Theory point of view, the options available as well as the payoffs are frequently more complicated than a simple 2×2 game. Steven J. Brams[5] has undertaken a detailed analysis of the Cuban Missile Crisis from the view point of Game Theory (more specifically using the 'Theory of Moves'—an offshoot of Game Theory).

It is also important to note that if an entity (say a 'soft state') would always swerve regardless of what the other player does, then the preferences do not match the situation outlined in this game. A different matrix may be needed to solve such a dilemma.

Volunteer's Dilemma

In this variation of the chicken game, a player notices that electricity has gone out for the entire neighbourhood. The repair team will come if someone calls to notify them. The volunteer who makes the call incurs some cost. This is, in effect, a multiplayer version of the chicken game, where 'volunteering' may be considered to represent swerving. The similarity with chicken is that if no one volunteers, the worst possible outcome is obtained and if any one person volunteers, the remaining benefit by not doing so. For *N* players there would be the same number of pure strategy equilibria, with one person volunteering in each. Of course, a different pure strategy equilibrium would be preferred by each player. All equilibria would be both Pareto optimal and Hicks optimal.

The War of Attrition and Other Timing Games

Another interesting variation of the chicken game can be if the outcome for each player, though dependant on the others, can also be independently arrived at based on the time factor. Let us again take two examples from popular culture. A variant of the game of Chicken as acted out in the movie *Rebel Without a Cause*, entails players racing towards a cliff. The winner is the player who jumps out of his car last. The pure-strategy space in the movie game is continuous while the payoffs are discontinuous at the cliff's edge, which makes the game more difficult to analyze mathematically. In the Hindi movie *Ghulam*, the protagonist runs towards an approaching train and must delay swerving for the longest time feasible. The game would be repeated with other participants being placed in a similar scenario. In this case again the payoff is discontinuous at the time of impact. The person who holds out the longest is the winner.

The War of Attrition is another game something like Chicken stretched out over time, where the game ends when one of the players picks Swerve. The payoffs for not ending the game are negative (penalties). Therefore, until the game ends both the players earn a negative amount of payoff per period of time. Thus no decision by either player is a mutually hurting stalemate. When one player does exit, the opponent earns a reward for outlasting him. The exact opposite of the War of Attrition game are pre-emption games like *Grab the Dollar*, in which the reward goes to the player who chooses the move which ends the game. The catch is that a cost is paid if both players choose that move, but no cost is incurred in the intervening period when neither player chooses it. In this game, a dollar is placed on the table between the two players, who each must decide whether (and when) to grab for it. If both decide to grab in the same discrete time period, both are fined one dollar. If one grabs and the other doesn't, the person getting the dollar gets to keep it.

Battle of the Sexes

In this game, a couple who had agreed to meet in the evening cannot recall if they had agreed to go to the opera or to a football match. The husband would prefer football while the wife would like to go to the opera. The catch is both would prefer to be together, rather than spend the evening at different places. It is assumed that they cannot communicate and each has to take a decision about the venue independently. The payoff matrix (figure 2.4) for this game has some interesting points.

	HUSBAND CHOOSES OPERA	HUSBAND CHOOSES FOOTBALL
WIFE CHOOSES OPERA	3,2	1,1
WIFE CHOOSES FOOTBALL	0,0	2,3

Figure 2.4: Battle of the Sexes

In this case each of the equilibria is deficient in some way, either being unfair to one (or both) of the player(s) or being inefficient. The players will also often miss to coordinate, leaving them with a payoff worse than going to a non-preferred destination. The worst scenario of course would be if they land up alone at a non-preferred event. Mathematically it can be shown that each player has an expected return of even lesser than the return one would receive from constantly going to one's less favoured event. If they could communicate, one resolution of the problem could be to use a toss of the coin (leading to what is called correlated equilibrium). Once the results of the toss are known, neither player has an incentive to deviate from the agreed upon choice as that would result in non-coordination and a lower payoff than simply adhering to plan of honouring the result of the toss. This results in perfect coordination at all times and is also fair in the sense that prior to the toss, the expected payoffs for both the players are exactly equal.

Other Games

There are a number of such games which have been proposed by various scholars. Once the basics are clear, it would be possible to apply the concepts to other situations as well.

NOTES

1. Phil Mellinger, 2002, accessed at http://www.gametheory.net/popular/reviews/GoodBadUgly.html on 18 January, 2012.
2. Kilgour, D. Marc; Brams, Stephen J., "The Truel", *Mathematics Magazine* 70 (5): 315–326 (December 1997).
3. Axelrod, Robert M., *The Evolution of Cooperation* (New York: Basic Books, 1984). As per other accounts The Prisoner's Dilemma was so named by Albert Tucker in an unpublished paper, although the particular 2-by-2 matrix was already well known. As recounted by Straffin (1980), Tucker was asked to give a talk on Game Theory to the psychology department at Stanford, and invented a story to go with the matrix.
4. IDSA comment: Game-theoretic Understanding of India-Pakistan Relations accessed at http://www.idsa.in/idsacomments/GametheoreticUnderstandingofIndiaPakistanRelations_agupta_160811 on 18 January 2012.
5. http://plus.maths.org/content/os/issue13/features/brams/index 160811 (Accessed 18 January 2012).

CHAPTER 3

Solving Games

Games may be solved by a variety of means including the use of mathematics or logical reasoning. A solution, or a solution concept, is a formal rule that may be used for predicting how the game will be played, and thereby the logical solution to the game.

Mathematical models do sometimes throw up seemingly illogical solutions or at least solutions which are not intuitive. Before going into the methods available for solving games, it may be useful to see whether the theoretical foundations hold valid for real life situations. Games must, of necessity, be well defined in a mathematical model. Further to provide the necessary traction to the modelling, they should normally be kept simple. Real life on the other hand is seldom without shades of grey. The theoretical results may therefore not fit in all situations. This does not mean that the theory should be disregarded. Even if not completely intuitive (some would say especially when not completely intuitive), the results may, at the very least throw up options which may have otherwise been overlooked.

Intuition or Mathematics?

As stated previously, it must be accepted that the use of mathematics does sometimes lead to apparently bizarre conclusions. It has however been asserted that "if the logic doesn't have holes in it, the conclusions are sound, even if they conflict with your intuition".[1] One game, incidentally named 'Pirates' may be a good starting point to discuss the supremacy of logic, even when apparently convoluted, over intuition.

The 'Pirate' game involves five rational pirates, A, B, C, D and E who must decide how to distribute the spoils amounting to 100 gold coins. The pirates have a strict order of seniority: A is senior to B, who is senior to C, and so on. The rules of distribution are:

- The senior-most pirate proposes a distribution of coins.
- All pirates, including the proposer, vote on whether to accept this distribution.
- If the proposed allocation is approved by a majority or a tie vote, it happens. If not, the proposer is thrown overboard from the pirate ship.
- The next senior-most pirate makes a new proposal and the procedure is repeated till a decision is voted as being acceptable.

The pirates cast their votes based on three factors. First of all, each pirate wants to survive. Second, given survival, each pirate wants to maximize the number of gold coins he receives. Third, each pirate would prefer to throw another overboard, if all other results would otherwise be equal. The pirates do not trust each other, and will neither make nor honor any promises between themselves apart from the main proposal.

It might, intuitively, be expected that Pirate A, making the first offer, will have to allocate very little, if any, to himself for fear of being out-voted and thrown overboard. The theoretical solution is however very different. The problem may be solved by backward induction.[2] The logic being that strategic decisions are all about "What will the next person do if I do this?"[3] The decisions that follow yours are therefore important. The ones that come before yours aren't, because you can't do anything about them anyway.[4] Bearing this in mind, the place to start is the point at which the game gets down to just two pirates, D and E. The senior-most pirate at this point is D, and his optimal decision is obvious: propose 100 gold pieces for himself and none for E. His own vote is 50 percent of the total, so the proposal wins. Now, working backwards, we add in pirate C. Pirate C knows (more importantly E knows that he knows—rules of the game being 'Common Knowledge') that if C's proposal is voted down, the game will proceed to the two-pirate stage and E will get nothing. So E will vote in favor of anything that C proposes, provided it yields him more than nothing.

C can therefore win with the following proposal: 99 for himself, 0 for D and 1 for E. Continuing the process, at the stage when four pirates remain, B considers what will happen if he is thrown overboard, when he makes his decision. To avoid being thrown overboard, he can simply offer 1 to D. Because he has the casting vote, the support only by D is sufficient. Thus he can propose B:99, C:0, D:1, E:0. Assuming A thinks through all these things, he can count on C and E's support for the following allocation, which is the final solution:[5]

A: 98 coins
B: 0 coins
C: 1 coin
D: 0 coins
E: 1 coin

It may be argued that 'real life' is not like that and 'theory' is therefore just that. A closer reflection would however suggest that theoretical foundations remain valid even though the rules may not be as well defined in any practical situation. For example, if the same were not true, in case of a factory where the workers are in a majority and may resort to a strike, it may be expected that their wages (or share of the profit) would be higher. The fact that pay is often fixed as per the organizational hierarchy is therefore worth reflecting on.

Sometimes backward induction can also give bizarre results due to the game itself being ill defined or due to the logic being incorrectly applied. A common example[6] often given to substantiate this is of the *Surprise Test*. A teacher tells her class that she intends to give them a test on some day in the day following week. The day when the test would be given is not specified and will come as a surprise. There is a pupil who decides to use backward induction to logically deduce the day of the test by working backward through the days of the coming week. He figures that if the test hasn't been given by the time school is over on Thursday, the teacher will then have no alternative but to set the test on Friday which is the last working of the school week. A test on Friday would, of course, then be apparent to all the students and would go against the teachers stated aim of giving a 'surprise'. If the test were to be given on Friday, no one would be surprised. It therefore logically follows that the teacher can't plan to give the test on Friday. This leaves only four days on which the test may be planned. Successive application of backward induction, following the same logic, it may be seen can eliminate Thursday as a possible day for the test. Once Thursday has been eliminated, and only three days are left, the student is next able to eliminate Wednesday. Following the same methodology, having eliminated all days of the week the student decides that the statement that a test would be planned for the coming week must be untrue. The logic seems flawless but closer examination would reveal the problem with the game setting. Even if the backward induction argument is taken to be correct, there must still be an inconsistency in the two statements made by the teacher, viz that the test would be given and that it would come as a surprise. Therefore at least one of them must be wrong. There is no logical reason to deduce that the incorrect statement is about the test being given, and not about the timing of the test coming as a surprise. The game, to be correctly framed, must be specified as the teacher choosing one of five days of the school week for the test and the student predicting which of the five days will be chosen. If the prediction about the day of the test is incorrect, the student will be taken by surprise. In other words, if they both pick a day at random, the student has 80 per cent chance of being surprised (only one out of the five possible days can be correct). This conclusion expects the student to predict in advance (before the moves are made or before the week starts) and thus does not permit him to predict that the test will be today, on each day of the week. In this alternate incorrect game previously solved by the student, his optimal strategy (were it to be permitted) would be to predict Monday on Monday, Tuesday on Tuesday, Wednesday on Wednesday, Thursday on Thursday, and Friday on Friday and thus ensure that he could never be surprised. Alternately he would reach the apparently paradoxical conclusion that there can be no test, as the teacher cannot possibly surprise them!

Options for 'Solving' Games

Before seeing some of the solution concepts *per se*, it must be acknowledged that a number of options are available for solving most problems. Some may be easier to apply in the given circumstances, whilst others may be more universally applicable. Mathematical solutions may be preferred by some people (which again can be obtained graphically or by the use of algebra), whilst others may prefer logical reasoning using nothing but plain simple English to obtain the same solution. There can therefore be no set rules for how games should be solved, but a generic understanding of the commonly used solution concepts will help understand how to tackle a given problem.

To emphasise the multiplicity of solutions, we will see an interesting puzzle (though not strictly a 'game' as previously defined, this puzzle could be converted into one). It is given that a hundred people are standing in a circle in positions numerically specified as 1 to 100. The person in the Number 1 position has a sword. As per the puzzle he is supposed to kill the next person (i.e. Number 2) and thereafter hand over the sword to the next-to-next (i.e. Number 3). The same procedure is to be continued till only one person survives. This simple puzzle is a variation of the Josephus Problem.[7] The question to be answered is which person would survive (by extension if you could choose a position before the game started, which position should you logically choose?).

It is possible to solve the problem by simply writing the numbers on a piece of paper and striking them out in sequence as per the given rules. In fact if the number of people were lesser, this may be the preferable solution. If the initial number was 10, for example, instead of 100, the answer could literally be on your fingertips. As the initial number gets larger, however, this approach becomes cumbersome and may also be prone to errors due to the repetitive, boring nature of the task of calculating. It must however be acknowledged that the puzzle can be solved even without application of any logic or mathematics, simply in a mechanical manner. Further, most people would find the logic becoming apparent to them as they set about this cumbersome task. In any case this is not a solution that rational people would contemplate.

We will next try and solve the problem logically. It is immediately apparent that Number 1 would kill 2, Number 3 would kill 4 and so on. The even numbers get killed and the odd numbers appear to be safe, at least in the first round. Further, it is apparent that as long as the total number of people we initially started with (100 in this case) is a multiple of 2, the sword will return to the person numbered 1 on completion of the first round as after Number 99 has killed Number 100, Number 1 would again be the next to receive the sword. At this stage the number of people remaining has been reduced to half but the logic outlined previously continues to hold good, albeit with the reworked sequence of 'even' or 'odd' positions. To further extend this logic, if the total number of people which we began with is a power of 2, the number of people remaining after each round would continue to be an even number. As long as this number is a multiple of 2, the last person standing would be Number 1. The largest power of 2 that is smaller than the given number 100 is 64 as the next power would be 128 which exceeds the number of people. As we have reasoned that a starting number that is a power of 2 does not materially affect the sequence as the sword will eventually come back to Number 1, we try to reduce the number of people (from 100) to a power of 2. To get to this number, we need to kill 100-64 = 36 people. Therefore, in the preliminary round itself, the first 36 people to be killed would be all even numbers up to 72 (i.e. Numbers 2,4,6,8,....72). After the 36th kill, the 73rd person would be handed over the sword. Starting with the 73rd person (if we consider him the new Number '1'), we could continue the game with the sword always returning to him after every round. Thus, he would ultimately be the lone survivor.

Next we observe that the same logic could be generalised to 'N' people instead of the given 100. The logic, when generalised to N people, results in the formula as given below:

Find the largest integer 'k' such that $2^k \leq N$

The answer is then: $2 * (N-2^k) + 1$

In the instant case, it is given that N = 100. The largest integer k is 6 (as $2^6=64$, and $2^7=128$). The answer therefore is:

$$2 * (N-2^6) + 1 = 73$$

Once again we get the same answer that the 73rd person will be left standing but we have moved to a stage where we can get the solution for any given similar problem, without dwelling into the logic. Indeed, a person ignorant of the underlying logic would be able to arrive at the correct answer if the 'magical' formula be known to him. He could, with the aid of a calculator if required, work out the solution to any given problem with similar rule sets.

We now take the problem one step further, in search of a more 'elegant' solution in addition to a 'generic' one. Suppose we were asked to write a simple computer algorithm for solving the problem. It is, of course, an easy task to write the programme if the mathematical formula is given, but let us for a moment think in 'binary' instead of 'decimals'. We observe the following:

- Removing the leading '1' in any binary number is equivalent to subtracting the largest power of 2.
- Adding a '0' at the end of a binary number is equivalent to multiplying it by 2.
- So, adding a '1' at the end is equivalent to multiplying by 2, and then adding 1.

Our formula entails subtracting the largest power of 2 (removing the leading '1'), multiplying by 2 and adding 1 (adding an extra trailing '1'). In effect if we could simply shift the leading digit to the last place we would readily have the answer (at least in binary terms). This also does away with the requirement to 'calculate the highest power of 2 lower that the given number' as regardless of the number, removing the leading digit means subtraction of whatever was the highest power of 2.

In the given problem:

N = 100 = 1100100 (in binary)

Shifting the first bit to the last: 1100100 → 100100$\underline{1}$

100100$\underline{1}$ = 73 (in decimal)

We once again reach the same answer, even though the 'procedure' adopted has no apparent direct link with underlying logic.

There are of course a number of other solutions available but the limited point sought to be made is that no one solution concept is the only solution, nor is any one of the possible solutions the best solution. It may also be seen that in the final analysis all the solutions flow from the same logic and are therefore interrelated even if not exactly the same.

Solution Concepts

The most commonly used solution concepts are structured around the concepts of dominance, equilibrium or optimality discussed previously. In case a game has *Strategic Dominance*, wherein the players are faced with a situation when one strategy is better than another strategy for any player, no matter how the opponent(s) may play, the solution is simple. Rationality demands that the other options be disregarded completely. A strategy that is strictly dominated implies that there is an alternate strategy available that has a higher payoff, regardless of the strategy that the other player may choose. In case of the Prisoner's Dilemma, to cooperate is strictly dominated by defect for both the players because each of them is always better off defecting, regardless of what the opponent does. Many simple games can be easily solved using dominance. To take another (very similar) game, consider two neighbours who share a common entrance to their flats. Each of them has an option of cleaning this common area and would obviously prefer the area to be clean. The effort required in cleaning the area means that they would each individually like to avoid the chore. We assume that if one cleans and the other does not, the person who does the cleaning gets a payoff of 2 whilst the person who took no effort and yet got a clean area gets a higher payoff of 6. If neither cleans, the payoff for them both is 3, since though the area remains dirty, neither has put in any effort. If player A decides to clean and get a payoff of 2, the player B gets a payoff of 6. If on the other hand player A had decided not to clean, the payoff for player B is 3 if not cleaning and 2 if cleaning. The option 'Don't Clean' thus gives a higher payoff for B regardless of what player A does. Similar reasoning by the other player implies that both players would choose not to clean and the area would remain dirty. The outcome 'Don't Clean' is therefore the dominant strategy equilibrium and will also be the solution for the game.

Nash Equilibria

Nash Equilibrium, as previously discussed, is a solution concept in which no player has anything to gain by changing only his/her own strategy unilaterally. The players are in equilibrium if a change in strategy by any one of them would lead that player to earn less as compared to retaining the current strategy, provided the other player maintains the previous strategy. This is possibly the most important solution concept in Game Theory. More importantly, John Nash proved that if we allow mixed strategies, then every game with a finite number of players in which each player can choose from finitely many pure strategies has at least one Nash Equilibrium.

There are a number of options available for finding the Nash Equilibrium in any given game. An easy way in a two player, normal form game is to 'circle' the maximum payoff in each row for the column player and each column for the row player. The cell with both the payoffs encircled represents a Nash Equilibrium. In other words if the first payoff number is the maximum of the column of the cell and if the second number is the maximum of the row of the cell—then the particular cell represents a Nash Equilibrium. This is also applicable in two player games in which players have more than two strategies and is an easy way of finding out the possible equilibria. Given any normal form game, we need to initially examine the first number to find the maximum of the column and, subsequently, check if the second member of the pair is the maximum of the row. If these conditions are met, the cell represents a Nash Equilibrium. All columns may be sequentially examined in this way so as to immediately discover all NE cells. The figure below (figure 3.1) depicts a two player game with three options each, where three NE have been established using this method (these have been highlighted for ease of reference). Theoretically an N×N matrix may have between 0 and N×N pure strategy Nash equilibria.

	Option A	Option B	Option C
Option A	0,0	12, 13	**5,15**
Option B	**40,24**	0,0	4,8
Option C	10,15	**15,25**	3,5

Figure 3.1: Nash Equilibria

The above rule however does not apply to the cases where mixed strategies are of interest. For mixed strategy games, the concept of stability can be applied to Nash Equilibria. A Nash Equilibrium for a mixed strategy game is said to be *stable* if a very small change (theoretically infinitesimal) in probabilities for one player leads to a situation where two conditions are valid:

- The player who did not change has no better strategy in the new circumstance.
- The player who did change is now playing with a strictly worse strategy.

Obviously in cases where the equilibrium is stable (where both conditions are met) the player with the small change in his mixed-strategy will prefer to return to the Nash Equilibrium. If the first condition is not valid, then the equilibrium is unstable. If the first condition holds but the second is not valid, there are likely to be an infinite number of optimal strategies for the player who changed. John Nash proved that the latter situation could not arise in a range of well-defined games.

Refinement

Very few games will have a dominant strategy equilibrium. Nonetheless, the concept of dominance can still be useful even while not providing a straight solution as in the Prisoner's Dilemma.

Most solution concepts for the more complex games will often result in more than one solution. We have previously seen examples of games which have more than one equilibrium. This can put any one of the solutions in doubt or, worse still, lead to no solution at all. Considering the game in stages may help in some situations where the solution is not apparent. *Refinement* is a process by which solutions may be narrowed down with each successive solution concept improving on the predecessor by eliminating equilibria which are unlikely or implausible.

To understand the concepts of refinement, we will see an example of a game called 'The Battle of the Bismarck Sea', set in the South Pacific in the Second World War. The two players are the Japanese Admiral Imamura and the American General Kenney. Admiral Imamura needs to transport Japanese troops from the port of Rabaul in New Britain across the Bismarck Sea to New Guinea. Kenney on the other hand wants to prevent him from doing so. The option available with him is to bomb the vessels carrying the troops during the transit. The options available to Imamura are between the shorter Northern route or to take a diversion through a longer Southern route to New Guinea. Kenney, who has limited assets to undertake the interception, must decide where to send his planes for reconnaissance. Once the ships are located they can be continuously tracked and bombed. If Kenney sends his planes to the wrong route he can subsequently redeploy them to the alternate route, but the time wasted in the fruitless search implies that the number of days of bombing is effectively reduced. The payoffs represent the number of days of bombing that would be feasible. With a longer route, more number of days are feasible even though the initial search is unproductive. The players have the same action set, {North, South}, but their payoffs are different as is shown below (figure 3.2).

		Imamura	
		North	South
Kenney	North	2, -2	2, -2
	South	1, -1	3, -3

Figure 3.2: The Battle of the Bismarck Sea

This is a zero sum game in that Imamura loses exactly what Kenney gains. Because of this feature, the payoffs could have equally been represented using just four numbers instead of eight. The 2-by-2 form with just four entries is termed a matrix game, while the equivalent table with eight entries (as has been shown) is a bi-matrix game.[8]

As may be seen, neither player has a dominant strategy (an option that gives a higher payoff regardless of what the opponent does). Their decision would be based on what they expect the adversary to do. Kenney would logically choose North if he thought Imamura would choose North, but South if he thought Imamura would choose likewise. Imamura, who wants to avoid the engagement, would choose North if he thought Kenney would choose South. The important difference (due to the varying length of the routes) now becomes apparent. If Imamura thought that that Kenney would choose North, he would be indifferent between actions he has to take as both of them give a payoff of 2.

This game does not have a dominant strategy equilibrium since one of the players (Kenney) does not have a dominant strategy. But we can still find a plausible equilibrium, using the concept of 'weak dominance'.

A strategy is said to be weakly dominated if there exists some other strategy for that player which is possibly better and never worse, yielding a higher payoff in some strategy profile and never yielding a lower payoff. Imamura's strategy of South is weakly dominated by the strategy North because his payoff from North is never smaller than his payoff from South, and it is greater if Kenney picks South. We can identify a weak dominance equilibrium as the strategy profile found by deleting all the weakly dominated strategies of each player.

We next observe that for Kenney neither strategy is even weakly dominated. Eliminating weakly dominated strategies will not therefore help much in the Battle of the Bismarck Sea. A further step, may help resolve the problem. Successive application of solution concepts may be gainfully employed to reach an *iterated dominance equilibrium.*[9] An iterated dominance equilibrium is a strategy profile found by deleting a weakly dominated strategy[10] from the strategy set of one of the players and then reconsidering the balance sub-game anew to calculate which of the remaining strategies are dominated. Iterated deletion of the next dominated strategy may be feasible until only one strategy remains for each player.

In the Battle of the Bismarck Sea, Kenney may reason as follows:

- Imamura wants to choose the opposite of what I do. Luckily he does not know what I will choose.
- Any decision he takes will not affect Imamura if I pick North.
- If I pick South, he can have a worse result if he too picks South.
- Imamura will pick North because it is weakly dominant.
- So I can eliminate "Imamura chooses South" from consideration.

Having thus deleted one column, Kenney has a strongly dominant strategy: he chooses North, which achieves payoffs strictly greater than South. The strategy profile (North, North) is therefore the iterated dominance equilibrium. The resultant solution (North, North) was also the actual outcome in 1943.

The following aspects need to be highlighted when using this approach to game solving:

- A dominant strategy equilibrium of every game may not exist. It is however, unique for any given game if it does exist as each player has at most one strategy whose payoff in any strategy profile is strictly higher than the payoff from any other strategy. Only one strategy profile can be formed out of dominant strategies, which is also the solution to the game.
- While, a strong iterated dominance equilibrium is unique if it exists, a weak iterated dominance equilibrium may not necessarily be so. The order in which strategies are deleted can also affect the final solution. If all the weakly dominated strategies are eliminated simultaneously at each round of elimination, the resulting equilibrium is unique, if it exists, but it is also possible that no strategy profile will remain at the end of elimination.

Theoretically, we can consider variations of the Battle of the Bismarck Sea game by modifying the sequence of play or the character of the information. If the game was sequential rather than simultaneous (assuming the intelligence about the moves is available to the adversary), two variations present themselves:

- If Kenney moved first (North, North) would remain an equilibrium, but (North, South) would also become one. The payoffs would be the same for both equilibria.
- If Imamura moved first, (North, North) would be the only equilibrium. Kenney's information set becomes either {Imamura moved North} or {Imamura moved South} after Imamura's decision, so Kenney's equilibrium strategy is specified as (North if Imamura moved North, South if Imamura moved South).

Other Solution Concepts

Use of Linear Programs, mini-max theorem and the like which are necessitated in games which do not have a dominant strategy are discussed in detail later. As would be apparent, not all games will have a unique solution or a dominant strategy. A pure strategy is one where the player would play the same strategy each time the game is played. There are many games where the players play a 'mixed strategy' wherein a random element is introduced in their choice, with there being a 'probability' that a certain option would be exercised. Again there are many methods available for solving such games. The most common amongst these is the solution based on the minimax theorem where a player tries to minimize the possible loss for a worst case (maximum loss) scenario.[11] The theorem was initially formulated for two-player zero-sum games (for scenarios where the moves are simultaneous as well as those with sequential moves) but has since been extended to more complex games and to general decision making in the presence of uncertainty. The theorem was first published in 1928[12] by John von Neumann. The importance of the theorem can be gauged by the fact that Neumann reportedly said that "As far as I can see, there could be no theory of games ... without that theorem .... I thought there was nothing worth publishing until the *Minimax Theorem* was proved".[13]

To familiarize with the basic concepts, we will first see a simple two-player zero-sum game called *Matching Pennies*. The two players each show a coin. Player A wins if both coins show the same face, that is, either both show 'heads' or both show 'tails'. Player B wins if they show different faces. It is obvious that the two players each have two strategies, heads and tails. The normal form depicted below shows all possible strategy combinations, assuming A to be the Column player.

		Player A	
		Heads	Tails
Player B	Heads	–1, 1	1, –1
	Tails	1, –1	–1, 1

Figure 3.3: Matching Pennies

If the game were to be played repeatedly, there is obviously no 'best' or dominant strategy. A player who always chooses heads for example is certain to lose as the opponent will not take long to catch on and alter his moves accordingly. The use of mixed strategies is therefore the logical choice in a game such as this, where the whole point is to keep the opponent guessing. Even without mathematics, it is apparent that the ideal solution is to randomize between the two options (heads and tails). If both players use this mixed strategy, the result will be a 'Nash Equilibrium'. Each player would win half the time, which is the best that both can do given the strategy choice of the opponent.

It is certain that rational players would never randomize between two pure strategies unless they are indifferent between them. In other words, if one strategy were better, logically the other strategy would never be played at all. If both strategies are equal, each may be played with a certain probability. Player A, for example may decide to play heads 70 percent of the time and tails for the remaining 30 percent. If this 70-30 is the selected mixed strategy, the worst thing that could happen is that the opponent can somehow guess this choice. For any rational opponent who guesses that the mixed strategy involves playing heads more than half the time, the rational counter strategy will be to play tails all the time. Player A will then lose more than half the time and the overall payoff will thus be negative. If on the other hand, the mixed strategy required playing tails more than half the time, the opponent who knows the mixed strategy will certainly play heads all the time. Player A will thereby again lose more than half the time and the overall payoff will again be negative. The best strategy is therefore to play heads and tails equally often, which guarantees a payoff of exactly zero. This is the basis of the maxi-mini strategy propounded by Von Neumann. This involves working out the worst payoff a player could get on average from each of the mixed strategies, and then to choose whichever strategy would maximize your payoff if this worst-case scenario were always realized.

Linear Programmes

In most cases the problem to be solved or the game to be played would involve optimization of returns with given limited resources. In their simplest form quite a few such problems can be represented as a linear program in the standard form consisting of the following three parts:

- A linear function to be maximized.
- Problem constraints.
- Non-negative variables.

To take an example, we assume a farmer holding a piece of agricultural land, say L km^2, which he has the option of planting with either wheat or barley or a combination of the two. The farmer's resources will permit him to invest in a limited amount of fertilizer, say F kilograms, and insecticide, say P kilograms. The requirement for every square kilometer of wheat is F_1 kilograms of fertilizer and P_1 kilograms of insecticide. Similarly that for every square kilometer of barley is F_2 kilograms of fertilizer, and P_2 kilograms of insecticide. If S_1 and S_2 are the respective selling prices of wheat and barley produced per square kilometer, then the optimum areas to be planted with each within the given constraints can be worked out. If we denote the area of land to be planted with wheat and barley by x_1 and x_2 respectively, then a solution for maximizing profit (by choosing optimal values for x_1 and x_2) can be calculated. The problem, when expressed as a linear programme in the standard form is stated as under:

Maximize: $S_1 x_1 + S_2 x_2$ (maximize the "objective function"—the profit in this case)

Subject to: $x_1 + x_2 \leq L$ (limit on total area)

$F_1 x_1 + F_2 x_2 \leq F$ (limit on fertilizer)

$P_1 x_1 + P_2 x_2 \leq P$ (limit on insecticide)

$x_1 \geq 0, x_2 \geq 0$ (Cannot plant a negative area).

The name 'linear programme' is indicative of the fact that the constraints when plotted on graphs are lines. When we graph these constraints and exclude areas that the inequalities indicate, we will get a feasible region that contains the solution. The corner point theorem says that if a maximum or minimum value exists, it will occur at the corner point of this feasible region.

Mini-max Theorem

Linear programmes can also be used to solve some games in Game Theory. The mini-max theorem states[14] that "for every two-person, zero-sum game with finitely many strategies, there exists a value V and a mixed strategy for each player, such that:

(a) Given player 2's strategy, the best payoff possible for player 1 is V, and
(b) Given player 1's strategy, the best payoff possible for player 2 is –V.

In other words, a solution can be found such that Player 1's strategy guarantees him at least a payoff of V regardless of Player 2's strategy. Since the game is assumed to be zero-sum and any gain by one player can only be at the expense of the other, Player 2 can guarantee himself a numerically equivalent payoff of –V. The name mini-max implies that each player minimizes the maximum payoff possible for the other and at the same time also minimizes his own maximum loss.

To take another common example, we can consider a simple two player game called 'Stone-Paper-Scissors'. The game involves two players simultaneously making a hand sign corresponding to one of the three options.[15] Stone can break the scissors, scissors can cut paper, and, paper can wrap the stone. So playing stone beats scissors, scissors beats paper, and paper in turn beats stone. When both persons play the same action, the game is drawn and none gets any points. The 'payoff' for a win is 1, with the opposing player getting a negative payoff (–1) for losing. The payoff is 0 for both the players in case of a draw. In Game Theory terms, this is a two-player zero-sum game which does not have a Nash Equilibrium. There is no pure strategy solution for this game and the optimal mixed solution is (1/3, 1/3, 1/3) or using each of the actions with equal probability. It can be proven that, no matter who makes the decision first, the payoff is same and the game is thus fair to both of the two players.

If we construct a payoff table for the game, the following matrix is obtained (only the payoffs for Player 2 are depicted, the payoffs for Player 1 would be the opposite):

	Player 1		
Player 2	Stone	Paper	Scissors
Stone	0	-1	+1
Paper	+1	0	-1
Scissors	-1	+1	0

Figure 3.4: Stone, Paper, Scissors

If the probability of Player 2 choosing the three options (Stone, Paper, Scissors) respectively is x_1, x_2 and x_3, the expected payoff of Player 2 is (x_2–x_3) when Player 1 plays Stone, (x_3–x_1) when Player 1 plays Paper and (x_1–x_2) when Player 1 plays Scissors. The best Player 1 can do is to achieve the minimum of these three values, namely

$$V = \min\ [(x_2 - x_3), (x_3 - x_1), (x_1 - x_2)]$$

So, Player 1 should use probabilities that minimize V. Similarly Player 2 should maximize the same. Therefore, when the Player 2 chooses a mixed strategy (x_1, x_2, x_3) then the optimal strategy can be found by solving the Linear Program.

$$\max z = v$$

s.t.

$$v = x_2 - x_3 \quad \text{(Stone constraint)}$$

$$v = -x_1 + x_3 \quad \text{(Paper constraint)}$$

$$v = x_1 - x_2 \quad \text{(Scissor constraint)}$$

$$x_1 + x_2 + x_3 = 1$$

$$x_1, x_2, x_3 \geq 0;$$

Similarly a value 'w' can be found from the perspective of the opposing player. The common numerical value of v and w is the **value** of the game.

The value of Stone, Paper, Scissors is $v' -1 = 0$

The optimal strategy for the row player is (1/3, 1/3, 1/3).

The optimal strategy for the column player is (1/3, 1/3, 1/3).

Von Neumann focussed on two-person, zero-sum games because a unique rational solution could be realized in each of them. It is paradoxical that his proof of this fact should be called the mini-max theorem, as the rational solution of a two-person, zero-sum game is actually for each player to apply the maxi-min principle. This involves first working out the worst payoff a player could get on average from each of the mixed strategies, and then to choose whichever strategy would maximize the payoff if this worst-case scenario was realized.

It would also be apparent that while some solution concepts are more convenient in certain situations, most concepts would be related and the same game can be solved by multiple methods. For example, if we take the case of backward induction, we are in effect discarding strategies that are dominated. Every time a group of choices is disregarded at any stage of the game while carrying out backward induction, effectively the equivalent pure strategies are being discarded. Any strategy that is considered amenable to being discarded at any stage would be the one that is dominated by a strategy which is the best choice to be made at the node in question. Except when two strategies always yield the same payoff (making the player indifferent to them), one strategy is dominated by another if it never yields a better payoff, no matter what strategies the other players may use thereby making some option(s) superfluous. Successively deleting dominated strategies in a game in its strategic form is effectively the same as backward induction. Further, it can be shown that in games of perfect information, backward induction will always leads to strategies that are not only a 'Nash Equilibrium' in the whole game, but also in all its subgames.[16] This is true even if they do not lie on the equilibrium path of an extensive form game.

Similarly, Von Neumann's mini-max theorem follows logically from Nash's proof that all finite games have at least one Nash Equilibrium, even though the two are unrelated. If we first determine, a Nash Equilibrium in a two-person, zero-sum game, the relation between the two will be apparent. For ease of reference we call the two players Adam and Eve. We, further term Eve's equilibrium strategy A, and Adam's equilibrium strategy B. The equilibrium payoffs will be called Eve's *value* and Adam's *value*. To go back to the game of Matching Pennies, the optimal choices A and B are the mixed strategy in which heads and tails are played with equal probability. Both the *values* are the zero payoff that each player will get (on average) if the game is played this way. Eve can never be certain of getting more than her *value* because Adam might always play B, to which her best reply is A. A moment's reflection will however show that she can be sure of getting at least her *value* by playing A because the best that Adam can do is to play B, which assures her of the *value*. We further consider that the best that Eve can do for herself in a zero-sum game is the same as the worst she can do to Adam. So her value is her maxi-min payoff. By extension A is one of her maxi-min strategies. By applying the same reasoning to Adam his *value* is his maxi-min payoff and B is one of his maxi-min strategies. In a zero sum game, the *values* of the two players will add up to zero. As we have seen the *values* correspond to the maxi-min payoffs, therefore the maxi-min will also sum to zero. In other words, neither of the players can get more than this maxi-min payoff unless the other gets less. When playing against a rational opponent, one can't improve on the maxi-min principle when playing a two-person, zero-sum game. This may be seen as Von Neumann's postulation, but is equally a logical conclusion from Nash's Theorem.

NOTES

1. Stewart, Ian. "A Puzzle for Pirates." *Scientific American*. 280.5 (1999): 98-99.
2. The concept of backward induction has been discussed in detail later in the chapter on dynamic games.
3. ibid.
4. This may seem contrary to the more 'accepted' method of decision making where the opponent's courses of actions are first listed and then an appreciation of own best course of action arrived at. Though the same would be necessary, the adversarial reaction would be as important and must be taken into account at the first stage itself rather than as a re-appreciation.
5. Other variants including for example, A:98, B:0, C:0, D:1, E:1 are not the same, as D would rather throw A overboard to get the same amount of gold from B. Variants also introduce an additional risk for the proponent, which by itself makes the variant less attractive.
6. Binmore, K.G., *Game Theory: A Very Short Introduction* (New York: Oxford University Press, 2007).
7. The Josephus Problem (also called the Josephus permutation) is named after Flavius Josephus, a Jewish historian living in the 1st century. There are many variations to the problem. The essence of the problem relates to a form of a counting-out game.

There are people standing in a circle waiting to be executed. The counting out begins at some point in the circle and proceeds around the circle in a fixed direction. In each step, a certain number of people are skipped and the next person is executed. The elimination proceeds around the circle (which is becoming smaller and smaller as the executed people are removed), until only the last person remains, who is given freedom. The task is to choose the place in the initial circle so that you are the last one remaining and so survive.

8. Games can be represented as matrix or bimatrix games even if they have more than two moves, provided that the number of moves is finite.
9. Sometimes the term is used to mean only deletion of strictly dominated strategies and sometimes to include deletion of weakly dominated strategies as well. The equilibrium concept used can also be called iterated dominant strategy equilibrium if the same be applicable. The game may be termed dominance solvable.
10. The difference between strong and weak dominance needs special mention. No rational player would use a strictly dominated strategy. Cases where an individual player is indifferent about the behaviour in an equilibrium due to weakly dominated strategies may however be debated. It is advisable in some situations not to completely rule out equilibria in which a player is indifferent about his actions.
11. This can alternatively be modeled as maximizing the minimum gain.
12. Published in German. Neumann, J. "Zur Theorie Der Gesellschaftsspiele." *Mathematische Annalen*. 100.1 (1928): 295-320.
13. Casti, J.L., *Five Golden Rules: Great Theories of 20th-Century Mathematics and Why They Matter* (New York: Wiley & Sons, 1996).
14. Osborne, Martin J, and Ariel Rubinstein, *A Course in Game Theory* (Cambridge, Mass: MIT Press, 1994).
15. A variation of the game is called 'Lady-Tiger-Hunter' and is very similar in execution. The tiger can eat the lady but be shot by the hunter. The hunter can fall prey to the ladies charm!
16. Reinhard Selten shared a Nobel Prize with John Nash for introducing this class of equilibria. These were initially termed perfect but are now termed subgame perfect. The concept is discussed in detail later in the book.

CHAPTER 4

Dynamic and Repeated Games

In the previous chapter we focused primarily on 'one-off' or static games though we did consider repeated iterations of games such as Stone-Paper-Scissors and Matching Pennies. Most situations in real life are seldom one-off events, but rather take place within the context of an ongoing interaction. This makes the games dynamic as the players can observe the moves of other players before making their responses. Repeated games and dynamic games need to be analysed separately. When a game is repeated it might be supposed that the players will progressively learn to co-ordinate their actions so as to avoid Pareto-inefficient outcomes even if the individual game is static or simultaneous. For example, while the decision to defect is the rational choice in the Prisoner's Dilemma and would likely be followed if the two players were certain that they would not encounter each other ever again, the players may think twice if they were partners in crime who could expect to work together subsequently. Reciprocity and expectations of fairness therefore play an important role in real life. Reciprocity can arise if the game requires the same players to play the same game over and over again.

A game can be dynamic for two reasons. If the players are able to observe the actions of other players before deciding upon their optimal response, the game is inherently dynamic in nature. In contrast, static games are ones where we can think of players making their moves simultaneously but even such a game is effectively dynamic if a one-off game is repeated a number of times. In such scenarios the players can observe the outcome of previous games before playing the subsequent games.

Some games involve players who, while being adversarial, are not strictly against each other (if that does not come as a paradox!). In other words, they would be quite willing to cooperate if the outcome was better for both (Win-Win) even though they have their own selfish interests which need to be protected or else there would be no game to begin with. Other games are strictly zero sum and there can be no cooperation—on the other hand there can be threats or implied threats. In either case the critical point is credibility. For a threat or promise to be credible it must be in that player's interest to carry it out at the appropriate time. If a threat or promise is not credible, then it would seem reasonable to suppose that it would be disregarded by the opponent. *Subgame perfection* applies this logic to dynamic games. In games of perfect and complete information, where players are not indifferent between various actions, backward induction generates a unique prediction. This prediction is the sub-game perfect Nash Equilibrium.

Promises or threats that are not credible may be considered *cheap talk.* Cheap talk refers to costless communication before the game proper begins. In Ranked Coordination, cheap talk permits each of the players to make clear the desirable outcome. In games such as Chicken, cheap talk is obviously useless, because each player will obviously announce the intention to never swerve. Cheap talk, as a coordinating device, can nonetheless be useful, especially in games like Battle of the Sexes. In such a game, coordination and conflict are combined. Without communication, the only symmetric equilibrium that is feasible is in mixed strategies, which is not optimal. If both players know that making inconsistent announcements will lead to a wasteful mixed-strategy outcome, then they would be willing to announce (provided announcements are permitted but are not enforceable) whether they will go to the opera or the football match. If there be multiple periods of announcements before the final decision, their chances of coming to an agreement are high. Thus communication can help reduce inefficiency even if the interests of the two players are in conflict.

Dynamic One-Off Games

An essential feature of all dynamic games is that the players can condition their optimal actions on the basis of what the opposing players have done in the past. This may, in some cases, enhance the options available.

Let us began by taking an example of a static one off game and next see how the situation changes in a dynamic context. For the static game we consider two firms (A and B) that individually have the option of entering a new market or staying out.[1] The decision of whether or not to enter is made difficult by the fact that the market is big enough to accommodate only one of the firms, but not both. Since some investment is necessary to enter that market, the limited market size implies that if both firms enter, they will both make a loss of 10 million. On the other hand, if only one firm enters, a profit of 50 million is expected with the passive firm making neither a profit nor a loss. The simple game, in the normal form is depicted below (figure 4.1).

		Firm B	
		Enter	Stay Out
Firm A	Enter	–10, –10	**50, 0**
	Stay Out	**0, 50**	0, 0

Figure 4.1: Entering a New Market

It can be seen that the two Nash Equilibria involve one of the firms entering and the other staying out. Further the mixed strategy Nash Equilibria can be computed by assigning the following probabilities

Probability of Firm A entering	Prob(Enter)_A
Probability of Firm B entering	Prob(Enter)_B
Probability of Firm A Staying Out	Prob(Stay Out)_A
Probability of Firm B Staying Out	Prob(Stay Out)_B

The expected profits for the Firm A if it decides to enter the market may be calculated by multiplying the payoffs with the probability with which Firms B's actions occur:

$$\text{Expected Profit Enter}_A = -10\ \{\text{Prob(Enter)}_B\} + 50\ \{\text{Prob(Stay Out)}_B\} \quad \text{(i)}$$

The expected profit if any firm decides to stay out is of course a straight forward zero. In case of an equilibrium, the two must be equal, therefore

$$-10\ \{\text{Prob(Enter)}_B\} + 50\ \{\text{Prob(Stay Out)}_B\} = 0$$

$\Rightarrow$ 50 {Prob(Stay Out)$_B$} = 10 {Prob(Enter)$_B$}

$\Rightarrow$ 5 {Prob(Stay Out)$_B$} = {Prob(Enter)$_B$}

$\Rightarrow$ 5 {Prob(Stay Out)$_B$} = 1– {Prob(Stay Out)$_B$}

$\Rightarrow$ 6 {Prob(Stay Out)$_B$} = 1

We therefore see that for the equilibrium, the probability of the Firm B staying out is 1/6 and of it entering is thus 5/6. Substituting these values to calculate the expected profit for Firm A {equation (i)}, we get

$$\text{Expected Profit Enter}_A = -10\ (5/6) + 50\ (1/6) = 0$$

The same result would also hold for the other firm. Thus the mixed strategy Nash Equilibrium for both the firms if they enter the market is zero, which is also the profit if they decide not to enter.

To understand how this can vary in a dynamic situation, we next examine a very similar two-period dynamic entry game.[2] Here again we consider two firms (C and D) that are considering whether or not to enter a new market. The other assumptions and conditions made previously continue to hold good, but to make this game dynamic we assume that firm D can observe whether firm C has entered the market before it decides what to do. To examine this game in the extensive form, we make the decision tree depicted below:

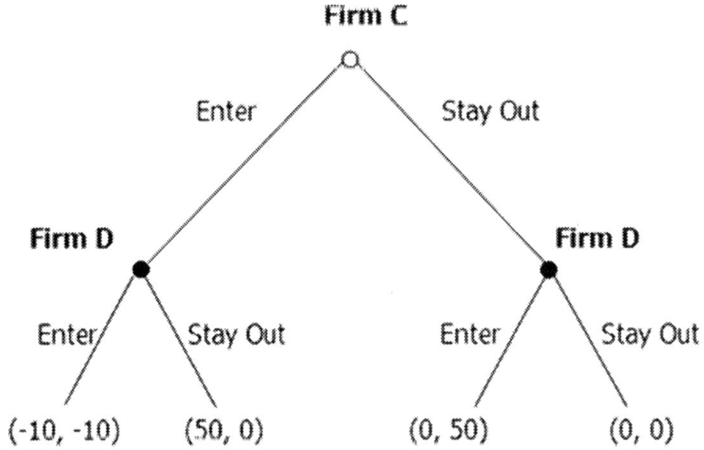

Figure 4.2: Dynamic Entry Game

In the first time-period, Firm C makes its decision, which can be observed by firm D which thereafter decides whether to enter or to stay out of the market in time-period 2.[3]

The fact that Firm D can observe the decision of Firm C before making its own decision obviously gives it an advantage. It may not be so apparent at this stage but being the first mover will also bestow some advantage to Firm C. If the two firms were to make their moves simultaneously, as had been considered in the previous game, each of them would have had only two strategies, namely to either enter or stay out of the market. However, because firm D initially observes firm C's decision it can make its decision conditional upon what firm C does. Since Firm C had two possible actions, as does firm D, it implies that firm D now has four (2 × 2) strategies as can also be seen from the extensive form representation. These may be listed as under:

Always Enter the market regardless of what Firm C does.
Stay Out of the market regardless of what Firm C does.
Do the Same as Firm C.
Do the Opposite of Firm C.

The two strategies of Firm C and the four strategies of Firm D can now easily be represented in the equivalent normal form game as depicted below (figure 4.3):

		Firm D			
		Always Enter	Always Stay Out	Same as C	Opposite of C
Firm C	Enter	–10, –10	**<u>50</u>, <u>0</u>**	–10, –10	**<u>50</u>, <u>0</u>**
	Stay Out	**<u>0</u>, <u>50</u>**	0, 0	0, 0	0, 50

Figure 4.3: Equivalent Normal Form Game

Having converted the extensive form game into a normal form game we can also easily determine the pure-strategy Nash Equilibria. When all players are playing their optimal strategies simultaneously, three such equilibria will be obtained as depicted.

Therefore, the dynamic entry game has three situations in which each firm is acting rationally given what it believes the opponent may do. To relate these directly to the threats or promises which the second player (Firm D) might make, we can consider Firm C as acting according to whether these are credible or not. The three equilibria can thus be explained as under:

- If Firm D threatens always to enter the market irrespective of what firm C does, Firm C will stay out of the market if it believes this threat. The outcome is {Stay Out, Always Enter} with the payoffs (0,50).
- If Firm D promises always to stay out of the market irrespective of what firm C does, Firm C will certainly enter the market if believes this promise to be true. The outcome is {Enter, Always Stay Out} with the payoffs (50,0). This promise is however not in the interest of Firm D and unless some binding agreements can be reached these would likely be not credible.
- If Firm D promises always to do the opposite of what firm C does, Firm C will again enter the market if it believes this promise. The outcome is {Enter, Opposite} with the payoffs again being (50,0).

In the first two Nash Equilibria Firm D's actions are not conditional on what the other firm does. In the third Nash Equilibrium firm D adopts a *conditional strategy*. A conditional strategy is where one player conditions his actions upon the actions of at least one other player in the game.

In each of the above equilibria Firm C can be seen to be acting rationally in accordance with its beliefs. What needs to be further examined, however, is whether these beliefs themselves are rational or not? To rephrase the question, one can ask if Firm C can rationally dismiss the threats or promises made by Firm D as mere bluff or cheap talk. Everything therefore will depend upon the credibility of the threat. In Game Theory a threat or promise is only credible if it is in that player's interest to carry it out at the appropriate time. Some of the statements by Firm D are not credible. For example, Firm D may threaten always to enter the market irrespective of what Firm C does, but the credibility of this threat is questionable. Once Firm C has already entered the market, Firm D's self interest would dictate staying out. The promise by Firm D to always stay out of the market is again not credible because if Firm C does not enter, then Firm D should certainly do so.

Sub-game Perfect Nash Equilibrium

A subgame perfect Nash Equilibrium (or simply subgame perfect equilibrium) is a refinement of a Nash Equilibrium often used in dynamic games. A strategy profile is said to be a subgame perfect equilibrium if it represents a Nash Equilibrium of every subgame of the original game. A subgame is formally defined as a smaller part of the whole game starting from any one node and continuing to the end of the entire game, with the qualification that no information set is subdivided. A subgame can therefore be considered a game in its own right. That 'smaller' game is not only a relevant part of the overall game but also one that may be played in the future. In other words this implies that if the players played any smaller game that consisted of only one part of the larger game and their behaviour represents a Nash Equilibrium of that smaller game, then their behaviour is a subgame perfect equilibrium of the larger game. It can be shown that every finite extensive game has a subgame perfect equilibrium.[4]

The above conclusion would be logical when examining all the Nash Equilibria obtained in the larger game. Assuming that players are rational, and that this is common knowledge, it seems reasonable to suppose that players would only believe credible statements (which may be threats or promises). This means that incredible threats or promises will not be believed. This further implies that incredible statements will have no effect on the adversary's behaviour as they will not be acted upon. In many dynamic games even though there are multiple Nash Equilibria, some of these equilibria may involve threats or promises that are not credible as it can be shown that to carry out these will not in the interests of the players making the statements. Simply put, the concept of subgame perfect Nash Equilibrium rules out such situations by the logic that a reasonable solution to a game cannot involve players believing (and thereby acting upon) incredible threats or promises. A subgame perfect Nash Equilibrium therefore requires that the predicted solution to a game be a Nash Equilibrium in every subgame. This amounts to requiring that a solution to a dynamic game must be a Nash Equilibrium in every subgame or that each player must act in his or her own self-interest in every period of the game.

We had previously seen that a common method for determining

subgame perfect equilibria in the case of a finite game could be through the use of backward induction. When one first considers the last actions of the game and determines which actions the final mover should take in each possible circumstance to maximize the utility, and further if this process is continued until one reaches the first move of the game then the strategies which remain are the set of all subgame perfect equilibria for finite-horizon extensive games of perfect information.[5] However, as seen previously, backward induction cannot be applied to games of imperfect or incomplete information because this entails cutting through non-singleton information sets. The set of subgame perfect equilibria for a given game is always a subset of the set of Nash Equilibria for that game. In some cases the sets can be identical. The sub-game perfect equilibrium necessarily satisfies the One-Shot deviation principle.[6]

We will again see the two firms to see how this equilibrium concept is practically applied. From the extensive form of this game it is apparent that there are two sub games. Each branch emanating from Firm D's decision nodes at time-period 2 can be considered a subgame. For the predicted solution to be a subgame perfect Nash Equilibrium it must comprise a Nash Equilibrium in each of these subgames. We will individually examine each of the Nash Equilibria identified for the entire game to identify the one(s) which (if any) is also a subgame perfect Nash equilibrium.

- Firm D threatens always to enter the market irrespective of what Firm C does. Out of the two subgames this strategy can be seen to be a Nash Equilibrium for only one. After Firm C enters the market, it is not in Firm D's interest to carry out the threat as the payoff for staying out is 0 whilst that for entering is –10. Firm D therefore will not enter regardless of the threat made earlier as the threat is not credible. Such a threat should logically not be believed by Firm C. This Nash Equilibrium is not subgame perfect.
- Firm D promises always to stay out of the market irrespective of what Firm C does. Once again this can be seen to be a Nash Equilibrium for only one of the two subgames. In this case, it is the optimal strategy in the subgame where Firm C has entered but not in the one when Firm C has stayed out. If Firm C stays out of the market, it is not in the interest of Firm D to keep its promise,

and so it will enter. This promise is therefore not credible, and so should again not be believed by firm C. Once more this Nash Equilibrium is not subgame perfect.

- The third Nash Equilibrium is characterized by Firm D promising to do the opposite of whatever Firm C does. This is a Nash Equilibrium for both subgames. If Firm C enters, it is optimal for Firm D to stay out, and if Firm C stays out, it is optimal for Firm D to enter. This is a credible promise because it is always in Firm D's interest to carry it out at the appropriate time in the future. This promise is therefore believable, and with this belief it is rational for Firm C to enter the market.

The only subgame perfect Nash Equilibrium for the game being discussed is therefore that Firm C will enter and Firm D, doing opposite of what it has observed Firm C doing, will logically stay out in its self-interest.

Repeated Games

Repeated games have many similarities with the one-off dynamic games just discussed. Each time a static game is repeated, it may be considered as a subgame of the larger game. In repeated games each of the independent one-off games is normally termed a *stage game*. Repeated games provide the players more opportunity to learn to coordinate their actions so as to avoid non optimum solutions as would otherwise be observed in games such as the Prisoners' Dilemma.

The number of times the stage game is repeated is important. Particularly important is the fact that the game may either be infinite or may have a finite number of interactions between the players. Whether or not the stage game has a unique Nash Equilibrium is also significant. Though, co-ordination may be possible in infinitely repeated games it is not possible in certain situations with finitely repeated games. This is true no matter how large the number of repetitions and is termed the 'paradox of backward induction'.[7]

To understand the issues involved with repeated games we will consider the situation between two competing pizza delivery firms in a particular locality. It is assumed that the two firms, between themselves, dominate the market and also that the customer base is equally shared between them. The customers do not have a particular liking for any of the brands and would therefore be indifferent to selecting between them if all other factors are the same. The firms know that reducing the price or offering a discount can increase sales *vis-a-vis* the competitor, all other things remaining the same. We assume that there are only two models of pricing (high or low) and that the total sales will also depend upon how much discount the rival firm is willing to give. In this game, both firms would be better off if they each maintain high pricing as they collectively dominate the market. If both firms decrease their prices simultaneously, then the market shares are unaffected, but both suffer a loss in overall profits. However, each firm has an incentive to unilaterally offer a discount in an attempt to increase their market share. If this game were played only once, then there is a unique 'Nash Equilibrium' with both adopting the low pricing model. This is a variation of the Prisoners' Dilemma where individual interest is against the collective good. Both firms would be better off if they both had high pricing.

The problem for the two firms is to co-ordinate their actions without the use of legally enforceable contracts. In the one-off game this would not be possible as there is a clear incentive to gain a greater market share and make up the loss in profit on each individual sale due to the enhanced sales. However, if the interaction between the firms is infinitely repeated, then it is possible for the two firms to co-ordinate their actions to reach the Pareto-optimal outcome. It is apparent that if both the firms reduce prices, leading to a 'price war', the market share would remain unchanged but the profit for both the firms will reduce due to the low pricing.

We have previously seen the concept of a conditional strategy wherein a player conditions actions based on the actions of at least one other player in the game. This is akin to saying Player A will adopt action X if player B does Y. Conditional strategy leads to other concepts such as punishment strategy and trigger strategy. If a player tries to deviate from the Pareto optimal outcome in an infinitely repeated game, he may be deterred from doing so by threats of punishment. If the prospect of punishment is sufficiently severe, and more importantly if the threat is sufficiently credible, players will be deterred from deviation. This can thus result in maintaining the Pareto optimal outcome indefinitely. Such an outcome may be termed as a *non-cooperative collusive outcome* in the sense that the players are in effect cooperating even though their interests are clashing. The necessary condition, namely of the threat being credible, implies that any punishment strategy will be effective in maintaining the co-operative solution only if it is part of a subgame perfect Nash Equilibrium for the entire game. For the pizza delivery firms, for example, each firm could start with high pricing model and maintain the same as long as the rival firm does so. If the rival firm shifts to the low pricing model then the logical thing for the first firm would be to follow the same. A threat that this pricing would be followed thereafter could be suitable punishment for the deviant firm as it ensures that the profits will dwindle in the long term. This particular form of a punishment strategy, where the actions of one player in a game causes other players permanently to switch to another course of action, is known as a trigger strategy. Trigger strategies can be effectively employed in many infinitely repeated games. A credible trigger strategy effectively rules out the possibility of ever returning to the Pareto optimal outcome, and is thus a severe enough punishment; provided the punishment itself as well as the threat of maintaining the punishment indefinitely, are both individually credible.

The practical use of the threat strategy can be observed even in other 'games' such as in the case of nuclear deterrence. A country that has a stated no-first-use policy is promising that good behaviour from the adversary would be rewarded. At the same time the threat of a massive retaliation in case of a deviation is assured in the form of an unacceptable punishment. Stating such a policy beforehand can make the threat of the punishment being inflicted credible.

Finitely Repeated Games

Whilst in an infinitely repeated game it may be feasible to maintain a non-cooperative collusive outcome, which is different from the Nash Equilibrium of the stage game, this may not always be the case in finitely repeated games.

Use of an appropriate strategy (such as the punishment strategy) would likely fail at the last stage of the game[8] as the players, knowing that there would be no further interaction, would behave as in the case of a stage game. There would therefore be no reason for deviating from the Nash Equilibrium. In other words, rational players using the principle of backward induction would deduce that in the last period the Nash Equilibrium will be played irrespective of what happens up to that period. Any threat of future punishment is therefore not credible. The threat can thus not induce a player to adopt any move other than the unique Nash Equilibrium in this final period. If we next consider the penultimate period, by logic of backward induction, again the Nash Equilibrium would be played. This argument can successively be applied to each preceding period until we reach the first period, where again the unique Nash Equilibrium would be played. The subgame perfect Nash Equilibrium for the entire game is therefore simply the Nash Equilibrium of the stage game played in every period.[9] This argument would seem to suggest that that a non-cooperative collusive outcome is not possible in a finitely repeated game, no matter how large the number of iterations, though this could be achieved in case of an infinitely repeated game quite easily.[10] Practically a number of means have been suggested by which the *paradox of backward induction* maybe overcome, including the introduction of bounded rationality, multiple Nash Equilibriums in the stage game, uncertainty about the future (and specifically about when the game may end), and uncertainty about the behaviour of other players in the game.

Reciprocity can also be tackled by means other than threats and promises. Strategics which involve a combination of these are also feasible. In an infinitely repeated game of Prisoner's Dilemma, for example, the use of the Tit-for-tat strategy is often advocated. As the name suggests, this involves the player initially seeking to cooperate in an effort to achieve the Pareto optimal solution and thereafter matching the last move of the adversary. The Tit-for-tat, unlike the trigger strategy, does not punish a deviation relentlessly. In other words deviant behaviour can be forgiven if the adversary shows a willingness to return to cooperation. The popularity of Tit-for-tat amongst social scientists stems from a competition[11] that showed this to be the best outcome. This is therefore sometimes thought to be the best paradigm for human behaviour across the board. Another strategy called Tit-for-two-tats is more forgiving and does not inflict punishment on the very first instance of deviation from the Pareto outcome. This has been shown to improve on the Tit-for-tat strategy under certain conditions.

NOTES

1. Romp, G., *Game Theory: Introduction and Applications* (Oxford [u.a].: Oxford University Press, 2005).
2. Ibid.
3. Ibid. In this extensive form game firm D's decision nodes are separate information sets. If they were in the same information set, they would be connected by a dashed line.)
4. Osborne, Martin J., *An Introduction to Game Theory* (New York: Oxford University Press, 2004).
5. Ibid.
6. The one-shot deviation principle is the principle of optimality of dynamic programming applied to Game Theory. It says that a strategy profile of a finite extensive-form game is a subgame perfect equilibrium if and only if there exist no profitable one-shot deviations. Ultimately, no player can profit from deviating from the strategy for one period and then reverting to the strategy. For further details refer to Tirole, Drew Fudenberg, Jean. *Game Theory* (6. printing. ed.), (Cambridge, MIT Press, 1991).
7. Romp, G., *Game Theory: Introduction and Applications* (England: Oxford University Press, 1997).
8. This is sometimes termed as the paradox of backward induction. Application of backward induction to finitely repeated game results in the sub-game perfect Nash equilibrium for the game being the same as the Nash equilibrium played in every time-period. This is true however large the number of repetitions.
9. Romp, G., *Game Theory: Introduction and Applications* (England: Oxford University Press, 1997).
10. In an infinitely repeated game, there is no endpoint from which to begin the logic of backward induction.
11. Robert Axelrod's Olympiad invited social scientists to submit computer programs to be matched against each other in the infinitely repeated game of Prisoner's Dilemma. The contestants submitted programs that implemented 63 of the infinite number of possible strategies for the game of which Tit-for-tat was found to be the most successful.

CHAPTER 5

Other Types of Games

Bargaining, pursuit, auctions etc. are all clearly forms of games. Since the rules are well defined and there are many applications of these games, a substantial amount of dedicated study has been conducted on solutions to the problems posed by these specific types of games.

For our purpose, ransom demands by pirates and negotiations with terrorist organisations are certainly variations of the basic bargaining game. Low Intensity Maritime Operations would involve pursuit of an adversary who attempts evasion. The terrorists would likely observe the actions taken by the security forces before choosing their own course of action. Existing studies into these situations therefore often refer to certain categories of games. We will see some of these so that their linkage with the practical application for the purpose envisaged is clear at a subsequent stage.

Bargaining Games

Bargaining may often be witnessed in economic transactions, such as buying property for example. If it is assumed that the economic transactions generate a surplus, the game addresses the question of dividing this surplus amongst the buyer and the seller. If the seller places the value of the object being put up for sale lesser than the value as determined by the potential buyer, the difference between the two valuations is the potential surplus. Each player therefore can bargain to get a greater share of this surplus. If there are mutual gains from the transaction it may be assumed that the negotiation process would be successful and the bargain will eventually be struck. On the other hand if the price being offered by the potential buyer is lower than the lowest that the seller is willing to accept, there will be no transaction and the negotiation will break down.

In its essence, therefore bargaining may be viewed as a problem of understanding how two players can cooperate when non-cooperation may lead to Pareto-inefficient result. Many games have multiple equilibria which may result in varying payoffs for each player. The players thus negotiate about which equilibrium to target. In a fair bargain it may be assumed that the final agreement is one that an impartial arbitrator would recommend. The basic Bargaining game (or more specifically the Nash bargaining game) is a simple two-player game used to model bargaining interactions. In the game, two players demand a portion of some desired commodity, which may be some amount of money or a piece of cake etc. If the total amount desired by both the players combined is less than that available, the players get their desired amount. If the total amount desired happens to be greater than that available, neither of the players gets anything. The *Nash bargaining solution* is the Pareto efficient solution to a Nash bargaining game. Variations of the game permit the bargaining to be mediated by an additional party, who is considered to be non strategic in its actions and has no personal interest in the outcome other than to ensure that it is fair to the strategic players. In other variations binding contracts may be permitted. In case of all negotiations, there is always the possibility of the negotiations breaking down. The value the players can expect to receive in the event of the negotiations breaking down is termed the *disagreement point.* Clearly the disagreement point is critical to the strength of negotiation and will directly affect the bargaining solution. Each player would rationally attempt to choose his disagreement point in order to maximize his bargaining position. Towards this objective, it is often advantageous to increase one's own disagreement payoff while harming the opponent's disagreement payoff.[1] Whilst each player bargains in an attempt to receive the best outcome sometimes it may be more effective if the person can convey that he is quite willing to walk away from the negotiation process.

Stackelberg Games

The Stackelberg Game based on the Stackelberg leadership model is a strategic game in economics in which the players are designated as 'leader' or 'follower'. The leader moves first and the follower(s) move sequentially. In the economic model the two compete on quantity. The problem is named after the German economist Heinrich Freiherr von Stackelberg who first described the model in *Market Structure and Equilibrium* (*Marktform und Gleichgewicht*) published in 1934. To examine the problem in Game Theory terms, the players of this game are a *leader* and a *follower*. In some ways this may be considered just another sequential move game in which the player moving second is permitted to observe the moves of the first, however there are many distinctive features of a Stackelberg Game. The peculiarities of the game include the following:

- The leader must be aware *ex ante* that the follower observes his action.
- Once the leader has made his move, he cannot undo it and is committed to that action.
- The follower must have no means of committing to a future non-Stackelberg follower action and the leader must know this.

In commercial ventures firms may engage in Stackelberg competition if one firm has some sort of advantage enabling it to move first. This may be possible if the leader was the incumbent monopoly of the industry and the follower is a new entrant. Moving first displays commitment in a most obvious manner. The follower deviating from the equilibrium can not only hurt himself but also the leader. Promises and threats studied earlier are therefore very much a part of any Stackelberg game. There is the underlying important assumption of perfect information in a Stackelberg game: the follower must observe the quantity chosen by the leader.

Cournot Competition

Cournot competition is an economic model used to describe an industry structure in which companies compete on the amount of output they will produce, which they decide on independently of each other. In Game Theory terms this is a simultaneous game. It is named after Antoine Augustin Cournot (1801–1877) who was inspired by observing competition in a spring water duopoly. The product is considered to be homogeneous or in other words there is no product differentiation between the outputs of various firms. There are multiple firms (the number of firms is fixed), all producing the same product. Cooperation between the firms is ruled out (there is no collusion). The firms compete in quantities and the decisions made by each individual firm can affect the price of the final good.

Pursuit-Evasion Games

Pursuit or evasion games, also sometimes referred to as *cops and robbers* deal with one player attempting to track down the other(s) in a given environment. The problem can be modelled both geometrically[2] (also called continuous pursuit evasion) and graphically[3] (also called discrete pursuit evasion). In the discrete graphical variant, the *cops* and *robbers* occupy nodes of a graph. The players move alternately and opt to stay put or move to the next adjacent node in the graph. If the pursuer occupies the same node as the one occupied by the evader, the evader is deemed to have been detected. In the continuous formulation the important difference is that the environment is modelled geometrically, typically taking the form of a Euclidean plane. Constraints of manoeuvrability, speed limitations etc. can be used to introduce variations. The games can be between two players (one each pursuer and evader) or multiple players (one pursuer attempting to detect multiple evaders, or many pursuers attempting to detect one evader or even many-on-many variants). The games obviously have direct military applications, especially in the maritime domain, where locating the adversary is one of the primary tasks. The concepts of reconnaissance (including Search and Patrol) are directly linked to pursuit evasion type of games. These games find direct application in many other military applications as well. In fact one of the initial applications of the pursuit problem was formulated at the RAND Corporation for missile guidance systems.[4]

NOTES

1. In this respect the disagreement may also be viewed as a as a threat.
2. See for example, Issacs, R., *Differential Games: A mathematical Theory with Applications to Warfare and Pursuit, Control and Optimisation* (New York: John Wiley & Sons, 1965).
3. Parsons, T.D., *Pursuit Evasion in a Graph, Theory and Applications of Graphs*, Springer-Verlag, pp 426-441.
4. Issacs, R., *Differential Games : A mathematical Theory with Applications to Warfare and Pursuit, Control and Optimisation* (New York: John Wiley & Sons, 1965).

SECTION TWO
REVIEW OF EXISTING STUDIES

APPLICATION OF GAME THEORY FOR
LIMO, ANTI-PIRACY AND COUNTER TERRORISM

CHAPTER 6

Review of Existing Studies

The use of Game Theory for combating terrorism and piracy has grown over the recent years. In the aftermath of 9/11, the volume of work related to terrorism has grown substantially as the realization that Game Theory may be an appropriate tool to model the interaction between the state and non-state actors has grown. It has been asserted that Game Theory is an appropriate tool for examining terrorism for a number of reasons.[1] These are summarized below:

- Strategic interactions between terrorists and a targeted government cannot be analyzed as though one side is passive as the actions are interdependent.
- Strategic interactions among rational actors, who are trying to act not only in their best interest but also according to how they think their adversaries will act and react can be used to model the interaction not only between the terrorists and the targeted government but equally to characterize the interactions amongst terrorists (e.g., between hard-liners and moderates or between two rival terrorist groups trying to get the same source of funding); between alternative targets (e.g., among two targeted governments or between two targets in the same country each of which is taking protective measures) or between two countries.
- Use of threats and promises to gain a strategic advantage is common in terrorist situations.
- The underlying rationality assumption of Game Theory, where a

player maximizes a goal subject to constraints is clearly applicable to terrorists and governments.

- Bargaining and negotiation is directly applicable to hostage negotiations and negotiations over terrorist demands.
- Lastly, uncertainty and learning in a strategic environment are relevant to all aspects of terrorism, in which the terrorists or the affected government or both are not completely informed.

The interaction between terrorist organisations and the counter-terror forces is obviously well suited for study from a Game Theory perspective. Each of these adversarial players is directly impacted by the action of the other. The options taken by the terrorists, historically, and how they have evolved in relation to actions by the defenders is interesting. An often quoted example is of the high number of airplane hijackings in the 1970s causing airports to increase the use of metal detectors. This made hijackings more difficult by increasing the costs for the terrorists.[2] An immediate result was observed in the increased number of kidnappings as the terrorists changed their strategy. Increased security to politicians and diplomats followed. This further resulted in the increase in the number of suicide bombings. In the Indian context, border fencing made infiltration from the land borders difficult, this may have encouraged the terrorists to look for other avenues and turn towards the sea.

Recent academic work has concentrated on studying not only the interaction between the government and the terrorists but also between two or more targeted countries, between terrorists and their supporters, between rival terrorist groups etc. Other, less-obvious player combinations include military and political wings of a terrorist group, terrorists and a state sponsor and even terrorists and the media.[3] A number of significant conclusions including some non-intuitive ones have thus been reached. The non intuitive conclusions are important as they permit the non-obvious insights to be uncovered which can thereafter be used to model the appropriate responses. More importantly these studies have paved way for further research into how the problems facing society can be tackled. Since the purpose of this book is to introduce the concept to a non-initiated reader, we shall first see some of the possible areas where Game Theory may be applied without going into the firm details at this stage. Some of the existing studies, the approaches adopted by them and the conclusions reached will first be examined.

Both cooperative and non-cooperative Game Theory can be used to study terrorism. We have focused on non-cooperative Game Theory till now and it would seem that the same, where the players have adversarial interests and cannot enforce agreements would be the more obvious choice for our purpose. It has however been asserted that cooperative Game Theory is "germane to understanding terrorist and government networks".[4] Games of interest in this field can include both static as well as dynamic games.

Bargaining games may have application to both anti-terrorism and negotiation of ransom demands by pirates. This has been studied in detail subsequently. In the first instance, it must be examined if the policy of non-negotiation with terrorists under all circumstances is feasible. A study by Harvey Lapan and Todd Sandler used a simple model to describe when, and if, a government should concede to terrorist demands.[5] The stated policy of the United States and some other countries is of non-negotiation and "no concessions to terrorists". Lapan and Sandler showed that this policy is optimal only if governments adhere to it and terrorists have incomplete information regarding government credibility.[6] The underlying logic of a stated no-negotiation policy, say in a hostage scenario, is that hostage takers would have nothing to gain and therefore the policy would effectively render hostage taking itself meaningless. This outcome "implicitly assumes that terrorists only gain from achieving their demands and that both sides are completely informed so that the subgame perfect equilibrium is to pledge not to concede. Obviously, something is incomplete about this logic, because terrorists continue to take hostages and even the staunchest advocates of the no-negotiation policy have reneged on their pledge".[7] The options available when preparing for a publically stated no-negotiation stance therefore need study from the policy perspective.

Countering terrorism can be done by either bolstering the defensive mechanisms or by adopting a more aggressive stance towards the terrorists by actively targeting them. In games where the state is considered to be one of the players, these two options provide the two most obvious options available to the player. Governments' anti-terrorism policies may thus be seen to be either proactive or reactive. Proactive policy involves aggressively targeting the terrorists in an attempt to eliminate their resources, infrastructure, and personnel. Reactive policy primarily concerns protective measures either to divert the attack or limit to its consequences. A pre-emptive strike against the terrorists or their state sponsors (for example, the Taliban in Afghanistan) is an extreme example of a proactive policy. A soft state on the other hand would take no proactive action at all. Of course, a mix of the two strategies would be the ideal and is often used by states to varying extents. The proactive action against the terrorists need not be restricted to military action (for example by targeting safe havens or training camps) but could also include other actions such as targeting their source of funding, etc. To study these options under the broad categories of 'defensive' and 'offensive' may be in order. These could equally be classified as reactive and proactive options.

The very aggressive posture by the United States in the aftermath of 9/11 is often contrasted with the more defensive policies of some other countries while attempting to answer whether proactive or defensive options may be the better choice. Proactive policies can also have the reverse effect, in that too severe a backlash may offer a fertile recruiting ground for even more terrorist activities. Rosendorff and Sandler[8] formulated a two-stage game where the government first chooses the level of being proactive and the terrorists subsequently choose the level of attack. The analysis indicates that spectacular events are encouraged by a proactive response that is either too meek or too strong. Choice of the proper intermediate response thus presents the government with a dilemma.

In another study, undertaken by Bandyopadhyay and Sandler[9] in 2007, a comparison between proactive and defensive measures was undertaken by "constructing a two-stage game in which each of two targeted countries chooses a proactive response in the first stage and a defensive effort in the second stage". The mix between these two policies depends upon on three key factors: the policies' cost comparisons, the countries' foreign interests, and the countries' targeting risks. The authors highlighted that studying proactive and defensive counter-terrorist responses in isolation provides an incomplete picture and concluded that prime-target countries with long borders and a high level of foreign direct investment are more likely to go proactively after a terrorist threat. Though the specific circumstances may make some countries lean more towards one or the other option, nonetheless, it is mostly agreed that a judicious mix of the two would normally be warranted to holistically tackle the problem.

Proactive Measures

Proactive measures are generally thought of as declaring a 'war on terror' as was done by the United States, though this terminology itself is questionable as terrorism is a tactic rather than an actor who can be targeted. Nonetheless, although the United States is the target of approximately 40% of all transnational terrorist attacks, virtually all of these attacks have occurred abroad in recent years with 9/11 being a noticeable exception.[10] Since a majority of the attacks on US interests have occurred oversees rather than in *Continental United States*, the US had little option but to go after the terrorists. The game-theoretic approach would however reveal a number of paradoxes in this approach. US over-deterrence and very proactive approach has resulted in attacks shifting abroad to locations where the US has "little authority to do anything about them".[11] Because a proactive policy of one state, if successful, can eliminate or substantially reduce the terrorist threat for all potential targets, there is also a tendency for other countries to free ride or rely on the efforts of others.

Proactive measures may not always be overt, and covert operations by Special Forces, where there can be deniability, have also been resorted to by some states to target the terrorist leadership. Targeted assassinations, as a method for controlling terrorism, has been studied[12] by Jacobson and Kaplan who showed that this is likely to be effective only if the level of terrorism is low to moderate. If the level of terrorism is already high these targeted hits can compound the problem by leading to a higher level of recruitment. Notwithstanding this backlash effect, the study shows that a government will still use this under certain circumstances.

Defensive Options

The first problem which would face all agencies tackling terrorism by defensive measures would be the question on how to allocate the counter terrorism resources amongst potential targets. Given the fact that the number of potential targets is large and the defensive resources available for their defence are limited, this problem becomes crucial.

Sandler and Lapan in a study undertaken in 1988[13] considered the potential targets as the players in this strategic interaction. The game therefore was played between two targets rather than between the terrorists and the targets. All other things being equal, the target which is the most vulnerable would be picked by the terrorist for an attack. Each target would therefore try and minimize the chances of an attack by strengthening its defenses and thereby minimizing its potential losses. The terrorist organization was considered to be a passive (or non-strategic) player. The conclusions of the study indicated that decentralized decision making may lead to overspending on domestic defences as each individual target would try and outdo the other. Every target is safe only as long as someone else is worse off than it. A centralized decision on the allocation of resources is therefore optimal. Interestingly, while it is assumed that more information about the preferred choice of targets would be helpful, the study concluded that increased information about the terrorists' preferences for targets may not lead to greater efficiency. Overspending on protection by each individual target may actually increase in such a scenario. The study further examined a globalization scenario where some countries may have similar interests at home and also abroad. In such a scenario, it was concluded, that deterrence is likely to be underprovided.

A very similar conclusion was reached in a study into the defensive problem faced by the ports and ships berthed in that port, each individually trying to protect itself. This study[14] by a team from the National Technical University of Athens, Greece applied Game Theory to a port security setting, where 'port' and 'ship' must tighten security to respond to a potential attack by terrorists. The level of their security measures and their associated costs were considered. Unlike totally diverse targets, the close juxtaposition of the two means that an attack can easily be diverted from one target to the other. Defence measures taken in coordination achieve the game's social optimum. In this case, the aim is to optimize the collective payoffs, while the players' individual actions may not be in their best private interests. Coordination between the opposing players thus becomes critical.

A game between the attacker and the defender was modeled by Bier,

Oliveros, and Samuelson[15] which involved the defender protecting multiple locations from a determined attacker. In this case it may be seen that the decision making has been centralized and there is therefore no strategic interplay between the targets themselves. Due to limited resources, however, each target would continue to have some vulnerability. The attacker has made the decision to attack and must therefore observe the defender's resource allocation so as to identify the most lucrative target. The preferences of the attacker are not known to the defender. Such a setting lends itself to both simultaneous and sequential games and the study cited examined each of these. In addition to reiterating the value of centralized allocation over decentralized allocation, the study reached an interesting conclusion about publicizing the defense allocations. It is normally assumed that the defences of any target must be kept secret or the attacker would gain an advantage. The study however showed that for an attacker who chooses a target based on the knowledge of the defence, the defender may actually be better off by making defensive actions public, especially if it is assumed that the more valuable targets would be better protected in a centralized decision making structure. It is only through this policy that the attacker can be directed towards the more vulnerable (due to being lesser protected) but less valuable targets. It was also shown that the defender is at least as well off in the sequential game relative to the simultaneous game and so the information asymmetry in the defender's favour may not have much of an impact. In equilibrium, some targets may be left unguarded.

Farrow (2007)[16] showed the gain from border control vis a vis defending all possible internal targets in terms of the marginal expected social cost avoided. For an optimal allocation between border control and protection of individual target sites, the aggregate marginal gain from border control must be equated to the common marginal gain from protecting individual sites. Some of the targets may not be individually defended. Strengthening coastal security, as seen in India in the aftermath of the 26/11 Mumbai attacks, can be seen from this perspective. The model adopted first studied a situation where targets are independent and subsequently studied a coordinated model.

It may not always be possible for the attackers to observe the defenses in detail before deciding on the targets. In some cases, the targets may be pre-selected based on other factors and a 'recce' of the defences subsequently undertaken. A situation where a defender must protect multiple potential targets from terrorist attacks and where the attacker cannot observe the defender's allocation among targets has been studied[17] by Powell. The expected losses at any given site depend on its vulnerability, threat (likelihood of attack), and consequences if attacked. The solution seeks to allocate resources among sites to minimize the terrorists' expected gain even though the game is not zero sum. The question of border defence *vis-a-vis* site defence is again examined with similar conclusions.

Protection of multiple locations from threats (which have been expanded to include not only terrorist threats but also natural disasters) was studied by Zhuang and Bier[18] (2007). This study therefore examines *safety* alongside *security*. Concepts of risk aversion, risk neutrality, and risk-loving behavior have been included and further both players are assumed to be aware of the opponent's preferences. The study concluded that the Attacker's efforts may rise with increases in defender's efforts, thereby not really resulting in an improvement in security. Safety, that is protecting against natural disasters, has fewer strategic concerns as the 'attacker' does not adapt its behavior or step up efforts in response to greater defensive investment. The study also showed that the defender is at least as well off in a sequential game as compared to a simultaneous game.

The underlying point which emerges in most studies is therefore that while potential targets invest in defenses to limit their vulnerability, the terrorists choose the specific target to attack based on observing these defences. If the targets act independently, then there will generally be over-deterrence as each target tries to divert the attack to another target. Solutions involving defence coordination will limit the overspending. Alternately, in case of given limited resources, the solution can be used to provide an optimum distribution.

Suicide Attacks

The subject of suicide attacks by terrorists has held some fascination for Game Theorists and the problem has been extensively studied. Of the face of it, the act itself may not be considered rational and therefore the application of Game Theory, which assumes rationality, is itself debatable. On the other hand it can be argued that even if the act itself may be motivated by irrationality the actions taken to achieve the ends are undertaken through logical reasoning. The suicide bomber would wish to cause an explosion at a place which can cause maximum devastation. It can also be argued that these are acts of self-sacrifice for furthering the group goals.[19] The rationality of suicide assumes that from the terrorists' viewpoint, the payoff from the attack, even when suicidal, is higher than the payoff from living. These perceived payoffs may be bolstered by a number of factors including, *inter alia*, rewards in the after-life;[20] payments to the family;[21] perceived altruistic gains,[22] and the prestige attached to the act of sacrifice as viewed by their peers.[23]

Suicide attacks are difficult to counter with defensive measures as standard target hardening may be ineffective where the opponent is willing to die to achieve the aims. It has been shown[24] and also practically observed that terrorists target the soft targets with conventional attacks whereas, the hard targets (which may be more lucrative and therefore better protected) are more susceptible to attacks through desperate means such as suicide bombings. Increased deterrence can thus reduce the overall number of terrorism incidents and yet increase the number of suicide attacks.[25]

Domestic Politics

A number of studies have also dwelt upon the influence that domestic anti-terrorism policies have on terrorism. Especially in the case of democracies, where public opinion is important, the voters can be considered as strategic players who influence decision making at the national level. The addition of voters, or public opinion on the government policies, to the interaction between policy makers and terrorists adds another important player to the interaction. The influence of domestic politics on counterterrorism has been studied by Siqueira and Sandler.[26] In the study, voters in two democratic countries that face a common terrorist threat choose policy makers during the elections. The policy makers must decide on the level at which proactive efforts would be taken. Due to the likelihood of a backlash if the efforts are too harsh, the proactive measures to reduce the threat from one terrorist group can increase the likelihood of its being attacked by another terrorist group. The second group is opposed to proactive countermeasures and will attack the country that is the most active in initiating countermeasures.

Though much of the Game Theory literature focuses on games where terrorists are interacting either with the government agencies or are considered to be non-strategic or passive players, some studies have also attempted to study the problem from the terrorists' viewpoint. A study[27] by Chelibik, for example, models a situation where terrorists are active players and nations, whether supporting or opposing terrorists, are passive players. The aim is to develop models which can provide insight into terrorist behavior which, in turn, can lead to useful counterterrorism policy recommendations. Here again the interaction can take many forms and can be modelled on rival terrorist groups, between hardliners and moderates, factionalism within a group etc.

Piracy

While the problem of terrorism from the Game Theory perspective has been well studied, academic examination of the problem of piracy has been comparatively more limited. This is surprising as, unlike terrorism which may be rooted in religion or ideology, piracy may be seen as being a purely economic transaction. This has been covered in detail subsequently. Some commentators have studied the economic business model[28] as applicable to piracy off the coast of Somalia, as distinct from a security perspective.

A study[29] by Carlos *et al* showed that the optimal response to a hijacking from the ship owners' perspective is always paying the ransom. This option allows the ship owner to keep the ship and also minimize the loss of lives. Clearly, this neglects the larger political implications of this action, especially in a repeated game. If no ship owner was ever willing to pay ransom or if some international agreement could have enforced compliance with a no-ransom rule, the problem would never have mushroomed to the extent that it did till it was again contained in the past few years. A number of factors such as proactive action by naval forces, presence of armed security guards on board ships, adherence to best management practices and international cooperation were all contributory factors towards the containment of the problem. Purely from the individual ship owner's perspective, the options would, of course, be very limited. These could include do nothing, use private protection (level may be varied) and use of alternate routes (make the much longer trip through the Cape of Good Hope, rather than the Suez Channel, thus avoiding the Somali coast and an eventual kidnapping). The study used a sequential defend-attack-defend model to show that the decision taken would depend on a number of factors including the level of risk aversion of the ship owner.

NOTES

1. Sandler, Todd, and Daniel G.A.M., "Terrorism & Game Theory," *Simulation & Gaming*. 34.3 (2003): 319-337.
2. See for example ibid, Sandler and Enders, "An Economic Perspective on Transnational Terrorism".
3. Todd Sandler and Kevin Siqueira, *Games and Terrorism- Recent Developments* (Sage Publications, 2008).
4. Ibid.
5. Lapan, Harvey E., and Todd Sandler, "To Bargain or Not to Bargain: That is the Question," *The American Economic Review*. 78.2 (1988): 16-21.
6. Ibid.
7. Sandler, Todd, and Daniel G.A.M., "Terrorism & Game Theory," *Simulation & Gaming* 34.3 (2003): 319-337
8. Rosendorff, B.P., and Todd Sandler, "Too Much of a Good Thing?: The Proactive Response Dilemma." *The Journal of Conflict Resolution : a Quarterly for Research Related to War and Peace* 48.5 (2004): 657-671
9. Sandler, Todd and Siqueira, Kevin, *Games and Terrorism- Recent Developments* (Sage Publications, 2008)
10. Sandler, Todd & Daniel G. Arce M., "Terrorism and Game Theory," *Simulation & Gaming*, Vol. 34 (3) September 2003.
11. Ibid.
12. Jacobson, D., & Kaplan, E.H., "Suicide bombings and targeted killings in (counter-) terror games," *Journal of Conflict Resolution,* 51(5), 2007 : 772-792.
13. Sandler, Todd, and Harvey E. Lapan, "The Calculus of Dissent: an Analysis of Terrorists' Choice of Targets." *Synthese*. 76.2 (1988): 245-261
14. Gkonis, Konstantinos G., Saraftis P., Harilaos N. and Ventikos, Nikolaos P., *Game Theory Contributions to Terrorism Analysis in Merchant Shipping: An Application to Port Security*, School of Naval Architecture and Marine Engineering, National Technical University of Athens.
15. Bier, V.M., Oliveros, S., & Samuelson, L., "Choosing What to Protect: Strategic Defensive Allocation against an Unknown Attacker." *Journal of Public Economic Theory*, 9.4 (2007): 563-587.
16. Farrow, S., "The Economics of Homeland Security Expenditures: Foundational

Expected Cost-Effectiveness Approaches." *Peace Research Abstracts Journal.* 44.4 (2007): 14-26.

17. Powell, R., "Defending against Terrorist Attacks with Limited Resources" , *American Political Science Review*, 101.3 (2007): 527-542.
18. Zhuang, J., & Bier, V.M., Balancing terrorism and natural disasters—Defensive strategy with endogenous attacker effort, *Operations Research*, 55(5), 2007: 976-991.
19. Enders, Walter, and Todd Sandler, *The Political Economy of Terrorism* (England: Cambridge University Press, 2006).
20. Berman, Eli, and David D. Laitin, *Hard Targets: Theory and Evidence on Suicide Attacks* (Cambridge, Mass: National Bureau of Economic Research, 2005).
21. Ibid.
22. Azam, Jean-Paul, "Suicide-bombing As Inter-Generational Investment," *Public Choice*, 122 (2005): 1-2.
23. Wintrobe, R., "Extremism, Suicide Terror, and Authoritarianism," *Public Choice*, 128 (2006): 169-195.
24. Berman, Eli, and David D. Laitin, *Hard Targets: Theory and Evidence on Suicide Attacks* (Cambridge, Mass: National Bureau of Economic Research, 2005).
25. Jain, S., & Mukand, S.W., "The Economics of High-visibility Terrorism", *European Journal of Political Economy*, 20(2), 2004: 479-494.
26. Siqueira, K., & Sandler, T., "Terrorist Backlash, Terrorism Mitigation, and Policy Delegation," *Journal of Public Economics*, 91(9), 2007: 1800-1815.
27. Chlebik, Kevin, Terrorism and Game Theory: "From the Terrorists' Point of View", *Pepperdine Policy Review*, Vol. III (2010): 15.
28. Carney, S. (2009), *An Economic Analysis of the Somali Pirate Business Model*, Wired.
29. Sevillano, Juan C., Insua D. Rios, and Jesus Rios, "Adversarial Risk Analysis: the Somali Pirates Case", *Decision Analysis* 9.2 (2012): 86-95.

CHAPTER 7

Use for Game Theory for Low Intensity Maritime Operations

The term 'Low Intensity Maritime Operations' (LIMO) is of relatively recent origin and there is some ambiguity as to the type of operations that come under its ambit. The term is mentioned only in passing in the Indian Naval Maritime Doctrine, the primary document that outlines the roles for the Indian Navy, and in the Maritime Strategy document. Therefore whilst some thinkers take the term to encompass only asymmetric warfare at sea, some others use a broader definition also encompassing the 'constabulary' role of the Navy under its ambit. 'Freedom to use the Seas' lists[1] under the broad heading 'Likely Scenarios for the Use of Military Force by the Indian Navy' the task "Ensure good order at sea which includes Low Intensity Maritime Operations (LIMO), to combat asymmetric warfare, poaching, piracy, and trafficking in arms/drugs". Admiral Arun Prakash (Retd) listing the roles of the Indian Navy in the NMF publication 'From the Crow's Nest'[2] classifies LIMO as a part of 'military' role, distinct from the 'constabulary' role of the navy at one place in the document and as a part of the constabulary role at another. Vijay Sakhuja on the other hand states that "Low Intensity Maritime Operations in the Indian context suggests application of force, in the maritime domain, against actors that engage in maritime terrorism, piracy, gun-running, drug smuggling, illegal fishing, poaching, marine pollution and criminal activities at sea that have the potential to disrupt order at sea. LIMO therefore involves use of force to an extent that it is short of conventional war against targets at sea".[3] For the purpose of this discussion, LIMO shall be taken to mean all naval operations in peace time against adversarial targets at sea.

There can be number of applications of Game Theory to LIMO. The primary requirement, from the planning perspective for such operations will be for 'scouting operations' comprising both 'search' and 'patrol'. It is assumed that once the target is detected, the naval forces would be quite capable of handling the ensuing operations which, by definition, are 'low intensity'. The problem therefore is of locating the target rather than of 'taking it on' once contact has been established. Lack of sufficient assets for detection—given the vast area of operations, and, producing 'actionable intelligence' from the huge amount of data available would be the primary requirements that need to be studied.

Search is defined as the systematic investigation of an area to detect the presence of or confirm the absence of an object thought to be in that area. Patrol on the other hand is systematic and continued surveillance along a line (known as a barrier line) to ensure that an object does not cross it undetected. Mathematically both operations are quite similar in their planning aspects. The problem arises when the resources available for the scouting mission are insufficient to cover the area of interest with the desired probability of detection. This is frequently the case in real life leading to what is known as a 'random' search where the available resources are optimally deployed and a lower probability of detection under the given constraints is accepted. While some consider the term 'random search' an oxymoron given that the search by definition is 'systematic' and thus opposed to the idea of randomness, others argue that 'randomization' itself may be systematically planned and executed. Game Theory has found many applications in this field. Deployed systems include versions of the ARMOR[4] (ARMOR-PROTECT for the US Coast Guard, ARMOR-TRUSTS for the Los Angeles Sheriff's Department, ARMOR-IRIS for the Federal Air Marshal's Service, ARMOR-LAX for the Los Angeles Airport Police and ARMOR-GUARDS for the Transportation Security Administration). The US Coast Guard PROTECT (Port Resilience Operational/Tactical Enforcement to Combat Terrorism) system website outlining the motivation[5] behind the program states that "Limited security resources prevent full security coverage at all times, which allows adversaries to observe and exploit patterns in selective patrolling or monitoring, e.g., they can plan actions avoiding existing patrols. Despite extensive outreach and coordination with other federal agencies, along with state, local authorities and industry there is simply not enough assets to maintain a 24/7 presence. This lack of assets extends to all 361 ports in the United States. Randomization of patrols and surveillance can be of help, but randomization must not mean haphazard operations. Randomization must still provide some quality-guaranteed security and very importantly, it must maximize deterrence".

Obviously having prior information about the adversary can be helpful. A fixed pattern of searching (or evading) is analogous to providing the adversary information that he can gainfully exploit. Since a pattern provides information to the adversary, it is only logical not to be predictive. Random patrol is therefore a good idea. This randomization must however not be haphazard.

There are a number of common games which can be directly applied to the problem of locating an adversary, including in cases where he may wish to remain hidden. An entire section of Game Theory called 'search games' is dedicated to one of the players attempting to locate the other. The 'Princess and the Monster' game is a well-known pursuit-evasion game played by two players in a defined region. The game was propounded by Rufus Isaacs in 1965 in his book Differential Games[6] as "the monster searches for the princess, the time required being the payoff. They are both in a totally dark room (of any shape), but they are each cognizant of its boundary. Capture means that the distance between the princess and the monster is within the capture radius, which is assumed to be small in comparison with the dimension of the room. The monster, supposed highly intelligent, moves at a known speed. We permit the princess full freedom of locomotion". Quite obviously with limited sweep of the sensors onboard, the naval vessel is a good analogy for the capture radius. The difference is that the payoff may, in practical terms, not often be in terms of remaining undetected for an unspecified amount of time which is sought to be maximized by the evader but in successfully reaching a target undetected or crossing an area without being caught. The fact that the exact choice of target amongst the many feasible may not be known to the law enforcement agencies complicates the problem.

Another interesting problem that can have direct correlation with maritime search operations is the well known 'Travelling Salesman Problem' (TSP). The problem was first formulated in 1930.[7] The problem statement itself is quite simple, 'Given a list of cities and the distances between each pair of cities, what is the shortest possible route that visits each city exactly once and returns to the origin city?' The solution can however be quite complex and there is no known algorithm that is guaranteed to solve every 'n' city problem in polynomial time (as a function of n). Brute force is completely impractical if the number of cities is large as the total number of possible tours when there are 'n' cities is (n–1)!/2. So, for instance, with 30 cities there are 4420880996869850977271808000000 possible tours. Calculating the length of each option to identify the shortest among them is obviously impractical in such a case. A number of methodologies for optimizing the route have been proposed and TSP is probably one of the most intensively studied problems in optimization. The problem also finds application in the context of planning aircraft searches where the aircraft must 'visit' contacts in order to classify them. In such cases the optimized search would involve visiting as many ships as possible in the shortest possible time. This therefore resembles the traditional TSP but has many significant differences[8]—notably the 'cities' (ships) are moving; time windows and precedence constraints are present; a total route length limit is enforced by aircraft endurance, and; the position of ships is usually only discovered as the route is flown. Again the time spent in each city, or the 'classification' time, makes the problem complex. It is often the time spent on a classifying a target rather than the travel between targets that is an issue in maritime surveillance. If the classification time is too small, then the scouting unit could be prone to misclassifying genuine neutrals as possible threats and may thus spend substantial time following benign individuals, increasing the possibility that the actual threat reaches its objective. The counter case where the threat is classified as benign is even worse since the object thus tagged may thereafter actually operate with impunity, especially in today's scenario where contact information (including classification) can be shared amongst various units. If the classification time is too large, then the threat may reach the objective or cross an area undetected before the scout detects and correctly classifies it.[9]

Patrolling Games

A simple patrolling game can be considered with the scout (Player A) trying to detect the presence of a target (Player B), where both the players have one ship/boat each.[10] The payoff is assigned with the conditional probability that side A catches B, when B decides to cross the area of operations undetected. This may be considered a zero sum game as both opponents have strictly opposite preferences. Player A strictly prefers catching the boat of Player B, and B prefers the opposite. The Area of Interest (AoI) may be divided into a number of strips, and Players A's pure strategy is to allocate the scout to a particular strip. Player B's pure strategy is to cross in one of strips. If there are N strips, the mixed strategy of A

$$a = (a_1, a_2, a_3....a_N)$$

where a_i is the probability the A will choose the pure strategy i.

and similarly for B

$$b = (b_1, b_2, b_3....b_N)$$

where b_j is the probability the B will choose the pure strategy j

For any pure strategy of A, say 'i' and any pure strategy for B, say 'j', we define P_{ij} as the probability that A will detect B. Player A would seek to maximize P_{ij} which may also be considered the payoff for A in a repeated game.

By Mini-max theorem, the value of the game, V can be calculated by

$$V = \underset{i}{\text{Max}}\ \underset{j}{\text{Min}}\ \Sigma\ a_i\ P_{ij}\ b_j = \underset{j}{\text{Min}}\ \underset{i}{\text{Max}}\ \Sigma\ a_i\ P_{ij}\ b_j$$

In other words for any given pure strategy of B, player A wants to employ a mixed strategy that maximizes V. To understand this as a linear problem, let us initially assume that B chooses the pure strategy b_1. Player A would then try to maximize V such that:

$$a_1\ P_{11} + a_2\ P_{21} + a_3\ P_{31} +...+ a_N\ P_{N1} \geq V \qquad \text{(i)}$$

If B had chosen the pure strategy b_2:

$$a_1\ P_{12} + a_2\ P_{22} + a_3\ P_{32} +...+ a_N\ P_{N2} \geq V \qquad \text{(ii)}$$

Subtracting (i) from (ii)

$$a_1 (P_{12} - P_{11}) + a_2 (P_{22} - P_{21}) + a_3 (P_{32} - P_{31}) + \ldots + a_N (P_{N2} - P_{N1}) \geq 0$$

Similarly considering each pure strategy for B, up to N would give the conditionality matrix

$$a_1 (P_{12} - P_{11}) + a_2 (P_{22} - P_{21}) + a_3 (P_{32} - P_{31}) + \ldots + a_N (P_{N2} - P_{N1}) \geq 0$$

$$a_1 (P_{13} - P_{11}) + a_2 (P_{23} - P_{21}) + a_3 (P_{32} - P_{31}) + \ldots + a_N (P_{N2} - P_{N1}) \geq 0$$

$$a_1 (P_{14} - P_{11}) + a_2 (P_{24} - P_{21}) + a_3 (P_{32} - P_{31}) + \ldots + a_N (P_{N2} - P_{N1}) \geq 0$$

and so on till

$$a_1 (P_{1N} - P_{11}) + a_2 (P_{2N} - P_{21}) + a_3 (P_{3N} - P_{31}) + \ldots + a_N (P_{NN} - P_{N1}) \geq 0$$

Given that $\Sigma\, a_i$ for (i = 1 to N) is equal to 1, the problem can be solved by linear programming. We assume 10 strips and assign the probability of detection as tabulated below:[11]

B Chooses→	1	2	3	4	5	6	7	8	9	10
A Chooses↓										
1	0.8	0.6	0.2							
2	0.6	0.8	0.6	0.2						
3	0.2	0.6	0.8	0.6	0.2					
4		0.2	0.6	0.8	0.6	0.1				
5			0.1	0.6	0.8	0.5	0.1			
6				0.1	0.6	0.8	0.5			
7						0.5	0.8	0.4		
8							0.4	0.8	0.3	
9								0.3	0.8	0.3
10									0.3	0.8

Notes: (1) Max Probability of detection (0.8) ensues when both players choose the same strip
(2) This reduces as the strip spacing between A and B increases
(3) Non-uniform reduction is assumed which reflects varying detection conditions based on characteristics of the area including density of traffic and geometry etc

Solving the game as a linear programme using the constraints matrix outlined above gives a solution of V = 0.1706. In other words A has a

17.06% chance of detecting B. It can also be shown that the optimal mixed strategy for A would be to select a pure strategy (particular strip) based on the solution with a probability that varies between 0 and 0.1719.

If the probability of detection is considered insufficient, A has two options. He could either increase the effectiveness of the scout or increase the number of scouts. It can be shown that the latter option yields a higher value of V. Doubling the capacity of the scout, for example by doubling the number of radars or lookouts (each individually having a probability of detection of 0.8 in the same strip) will obviously not double the overall value of the game.

The revised probability of detection P* is given by

$$P^* = 1 - \{(1-0.8)\ (1-0.8)\} = 0.96$$

Though that is a substantial increase in capability, reworking the value of V gives only a marginal increase in the value of the game. On the other hand giving two scouts enhances the probability of detection substantially (but still not double the initial value). The result, of course, is apparent even intuitively without the use of mathematics. If we have a larger area than we can search, two lesser capable scouts would be preferable to one more capable one.

To extend the problem, we next take a scenario where the two scouts have differing capabilities and must divide the total area into a sector assigned to each.[12] The question attempted to be answered is, can an optimum distribution of the area be arrived at so as to optimize the search much like the farmer optimized his profits for wheat and barley. We can game this as a variation of the princess and the monster where the princess does not have the freedom of movement. This assumption would be valid only for some scenarios[13] but is assumed to be true for the sake of simplicity. We further assume a Probability of Detection (PoD) = 1 when the 'princess' is in the 'capture radius' and PoD = 0 otherwise. The capture radius equals the sweep width of the scout and is denoted by 'W'. The scout has a 'patrol speed' 'V' and 'Time on Task' denoted by 'T'. For ease of calculations, if V or T is sufficiently large, we can assume the area (a) covered in time T[14] to be denoted by:

$$a = WVT$$

To maintain randomness we assume that the searcher can begin anywhere in the Area of Operations (A) and that neither of the players is aware of the others actions. If the entire area (A) could be covered, the PoD would obviously be 1. If only half the area can be covered, the PoD will be 0.5. If only a small area 'a' can be covered then the probability is the ratio of the area that can be searched to the total area.

$$PoD \mid A = a/A$$

For two scouts with different capabilities, the area that can be searched is:

$$a_1 = W_1V_1T_1$$

and

$$a_2 = W_2V_2T_2$$

The problem being examined involves dividing the total area A into two parts A_1 and A_2 such that they may be optimally assigned to the two scouts to maximize the combined PoD. Since the target may be anywhere, the probability of it being in A_1 and A_2 is denoted by $P(A_1)$ and $P(A_2)$ respectively.

$$P(A_i) = A_i \, / \, (A_1 + A_2)$$

To calculate the probability of detection

$$PoD \mid A_1 = a_1/A_1 \text{ and } PoD \mid A_2 = a_2/A_2$$

If the combined probability of detection is taken as P then

$$\begin{aligned} P &= (PoD \mid A_1)^* \, P(A_1) + (PoD \mid A_2)^* \, P(A_2) \\ &= (a_1/A_1) * P(A_1) + (a_2/A_2)^* \, P(A_2) \\ &= (a_1/A_1) * \{A_1 \, / \, (A_1 + A_2)\} + (a_2/A_2)^* \, \{A_2 \, / \, (A_1 + A_2)\} \\ &= \{a_1/(A_1 + A_2)\} + \{a_2/ \, (A_1 + A_2)\} \\ &= (a_1 + a_2) \, / \, (A_1 + A_2) \\ &= (a_1 + a_2) \, / \, A \end{aligned}$$

This non-intuitive result suggests that partitioning of an area does not matter as the terms A_1 and A_2 have no bearing on the result. This would hold good as long as there is no overlap between the areas.

To study the problem further we view it as a two player game where the Player A (with two scouts) is trying to detect the presence of a single target (Player B). Let us assume that the partitioning of the area is as before and is known to both the players. Player B has the option of locating in either A_1 or A_2. Player A has the option of deploying the assets in either of the areas. We assume that Scout 1 has better capability than Scout 2 (implying thereby that $a_1 \geq a_2$) and that area A_1 is larger than A_2. Placement of both scouts in one sub-area is not ruled out and both assets may simultaneously be present in the same sub area giving rise to the game matrix below.

Player 2 (Searcher)	Player 1 (Evader)	
	Area A_1	Area A_2
Option1	Scout 1	Scout 2
Option2	Scout 2	Scout 1
Option3	Scout 1 & 2	No Scout
Option4	No Scout	Scout 1 & 2

It is important to note that even when the scouts are both placed in the same sub area, there is no overlap of search areas as the area 'A' is much larger than the search capacity 'a'. Therefore the only restriction in partitioning is that either of the sub areas be at least as large as the sum of the search capacities:

Therefore $A_1 \geq (a_1 + a_2)$ and $A_2 \geq (a_1 + a_2)$

If both the players selected the first option, taking the PoD as the payoff, the payoff for the searcher would be (a_1/A_1). Similarly if evader selected Area A_2 while the searcher took option 4, the payoff would be $\{(a_1+a_2)/A_2\}$. The payoff matrix accordingly becomes:

Player 2 (Searcher)	Player 1 (Evader)	
	Area A_1	Area A_2
Option1	(a_1/A_1)	(a_2/A_2)
Option2	(a_2/A_1)	(a_1/A_2)
Option3	$\{(a_1+ a_2)/ A_1\}$	0
Option4	0	$\{(a_1+ a_2)/ A_2\}$

For convenience, we replace the terms as shown below:

Player 2 (Searcher)	Player 1 (Evader)	
	Area A_1	Area A_2
Option1	W	x
Option2	Y	z
Option3	$w+y$	0
Option4	0	$x+z$

We assume that $(a_1/A_1) = (a_2/A_2)$[15]

To establish the hierarchy of the terms, consider

Since $(a_1/A_1) \geq (a_2/A_2)$	$w \geq x$
Since $a_1 > a_2$	$w > y$ and $z > x$
Since $A_1 > A_2$	$z > w$ and $x > y$
$z > w$ *and* $x > y$	$(x+z) > (w+y)$

We therefore get the hierarchy:

$$z > w > x > y$$

Which gives the intuitive result that the more capable the unit or smaller the area, the greater the probability of detection, as summarized below:

$$(a_1/A_2) > (a_1/A_1) > (a_2/A_2) > (a_2/A_1)$$

and

$$\{(a_1 + a_2)/A_2\} > \{(a_1 + a_2)/A_1\}$$

It is not feasible to establish the hierarchical relationship between the terms (a_1/A_1) and $\{(a_1+ a_2)/A_2\}$. We will take one example of each of the cases. In the first example we take A= 100 where the area is divided into two sub areas such that A_1= 60 and A_2= 40, we further assume the search capacity a = 20 is based on individual search capacities a_1 = 15 and a_2 = 5. The payoff matrix becomes:

Player 1 (Searcher)	Player 1 (Evader)	
	Area A_1	Area A_2
Option1	15/60= 0.25	5/40= 0.125
Option2	5/60= 0.083	15/40= 0.375
Option3	20/60= 0.33	0
Option4	0	20/40= 0.5

For ease of visualization we can multiply each of the terms by a factor of 24 to obtain:

Player 2 (Searcher)	Player 1 (Evader)	
	Area A_1	Area A_2
Option1	6	3
Option2	2	9
Option3	8	0
Option4	0	12

Similarly it can be shown that if the area was 270 (divided into 150 and 120) and search capacity 100 (60 and 40 respectively for the two scouts), the payoff matrix (scaled for a factor of 30) can be formulated as is depicted below:

Player 2 (Searcher)	Player 1 (Evader)	
	Area A_1	Area A_2
Option1	12	10
Option2	8	15
Option3	20	0
Option4	0	25

The games may be solved graphically by plotting the four strategies of the searcher as a function of the evader's strategies. In both the cases (Figure 7.1 and Figure 7.2) the four lines intersect at a point, which is the value of the game. It may be noted that in each case the value of the game remains the ratio of the search capacity to the total area. Therefore, if the searcher plays his strategies optimally he would achieve a probability of detection equal to the value of the game. Further choices may be made based on the circumstances or by the expectation-variance principle.

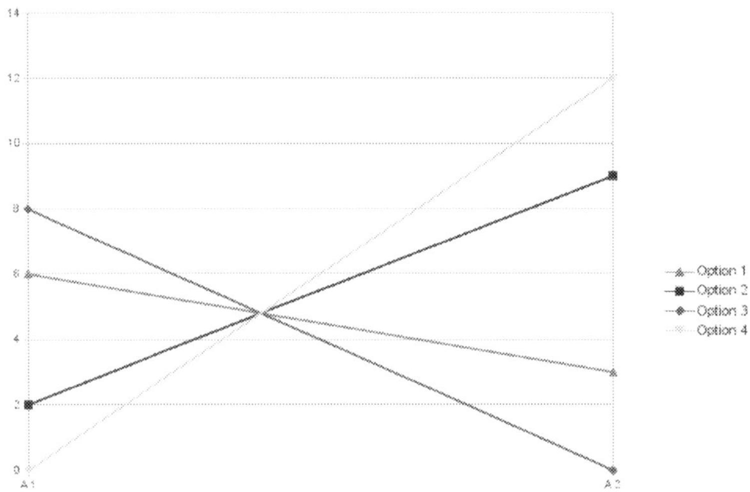

Figure 7.1

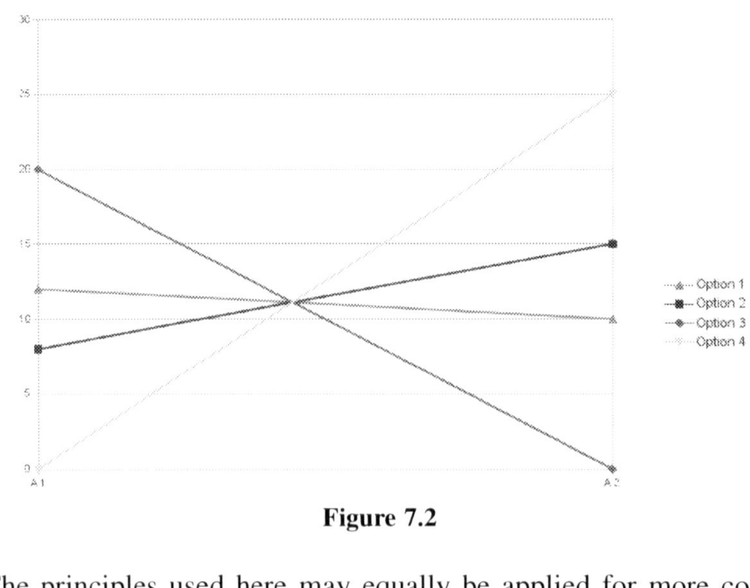

Figure 7.2

The principles used here may equally be applied for more complex games with more options to each player or by permitting the evader movement.[16] These can get fairly complex. For example just an addition of one sub area would lead to 84 games to analyse, even with only two scouts. Coverage factor greater than 1 is another interesting variation.

Implications and Policy Recommendations

The need to clearly and unambiguously define the term LIMO is important from the force structuring perspective. If LIMO will continue to be an important role for the Navy, then the apparently intuitive result that two lesser capable units give a higher payoff than one more capable one would need to be factored. Using a destroyer for LIMO may be not only less productive but also gross underutilization of the platform. Since this is not the only role, and the same platform may have to undertake multiple roles, a study into the right mix of assets may be necessary.

For the patrol game we had used the PoD as the payoff with the caveat that the payoff is assigned with the conditional probability that side A catches B, when B decides to cross the area of operations undetected. Using the reverse may be meaningless as B has the choice not to cross the area at all and can choose the time for the attack. This is often stated, even in other security problems, as the dilemma that the attacker has to 'get lucky' only once while the defender has to be lucky 24/7. If the cost of operations is taken as the payoff, the defending side would always 'lose' as even one instance of a successful attack would force constant allocation of resources to prevent a recurrence—and thereby entail a cost.

The allocation of areas or even overlap of areas between diverse agencies does not impact the PoD. Therefore any debate on optimal allocation of coastal security zones between various agencies may be irrelevant. It is however important to note that the conclusion is true only when there is no overlap of the actual search. This makes a strong case of coordination of patrols between diverse agencies.

Given that the probability of detection is a function solely of the size of the area of operation and the search capabilities, an analysis of the vast area to be covered in the Indian context may reveal the negligibly small payoff that may be feasible even with intense patrolling. The question that the policy makers need to answer is regarding quantification of the diminishing probability of detection and the threshold beyond which any patrol may be considered not commensurate with the efforts. If such a study is undertaken, alternative measures to protect the assets may need to be considered. This can include the following:

- Increasing attention to intelligence gathering so as to predict the likely time/place of the intrusion. Knowledge about the adversary's intentions can make the game less random and more predictable. The farmer could optimize his profits for wheat and barley and the extent of the area made a difference in the profits. That result was obtained because the yields for each area had different payoffs, unlike in the case of completely random behavior of the adversary. If even limited information about the adversary is available, the probability of a higher payoff in terms of sub-area deployment or time of operations will have a substantial increase in the PoD. This need not be what is often called 'actionable intelligence' or specifics of adversary deployment. Even a higher known (or assumed) probability of a particular option being exercised by the evader can substantially enhance the payoffs for the searcher.
- Reducing the area of interest by accepting detection of the adversary closer to own shores, even though early detection may usually be the preferred option may have to be considered. In conventional warfare, a weaker force usually adopts a 'citadel defence' concept where it waits for the adversary to arrive near the weapon release line rather than going out to look for him. In terms of LIMO, the conventionally stronger force may in effect be considered weaker due to the constraints outlined above. Strengthening fixed shore based surveillance may be the better option in such cases rather than patrolling. An optimal mix would be the ideal solution.
- Most planners would baulk at the idea of providing hundreds of operator displays at a shore based chain of radar stations and employing suitable manpower for monitoring and analysis. It is for consideration that even two ships on patrol already employ hundreds of people. The fact that out of the hundreds of crew onboard the ship, only a handful would be engaged in actual monitoring while others are engaged in routine 'running' of the ship; that the over-exploitation of the ship in terms of machinery exploitation and resulting increase in maintenance costs; and the cost of the fuel

itself may outweigh the cost of multiple radar displays ashore and of employing many analysts. The fact that the probability of detection would also be higher in the in the latter case should logically tip the scales in favour of this approach.

An appropriate agency may be considered for being tasked with developing suitable software for systematic randomization so that 'random' patrols are not haphazard. This could be done on lines of the ARMOR series of initiatives in the United States.

NOTES

1. *Freedom to Use the Seas: India's Maritime Military Strategy*, Integrated Headquarters Ministry of Defence (Navy), 2007, p. 60.
2. Prakash, Arun, *From the Crow's Nest: A Compendium of Speeches and Writings on Maritime and Other Issues*, (New Delhi: Lancer Publishers, 2007), pp 53 and 105.
3. Sakhuja, Vijay, *Asymmetric Warfare and Low Intensity Maritime Operations: Challenges for Indian Navy* (New Delhi: Observer Research Foundation, 2006).
4. ARMOR is an acronym for Assisted Randomized Motoring Over Routes.
5. http://teamcore.usc.edu/projects/coastguard/ accessed 12 Jan 2014.
6. Isaacs, Rufus, *Differential Games: A Mathematical Theory with Applications to Warfare and Pursuit, Control and Optimization* (New York: Wiley, 1965), pp 349-350.
7. The problem was first defined in the 1800s by the Irish mathematician W.R. Hamilton and the British mathematician Thomas Kirkman. It was formulated as a mathematical problem only in 1930 by Karl Menger.
8. Marlow, D.O., Kilby, P. and Mercer, G.N., *The Travelling Salesman Problem in Maritime Surveillance – Techniques, Algorithms and Analysis.*
9. For a detailed study of the classification time problem see Gaver, D.P., K.D. Glazebrook, and P.A Jacobs, "Search for a Malevolent Needle in a Benign Haystack," *International Series in Operations Research and Management Science*, 128 (2009): 125-146.
10. Based on the work of Nisgav Yair. See Nisgav, Yair, *Some Game Theory Models for Allocating Forces in a Narrow Strait against Hostile Activity,* Naval Postgraduate School, Monterey, CA, United States (1973).
11. Ibid.
12. Based on the work by Clark Wallis Pritchett. See Pritchett, Clark W., *A Game Theory Approach to Search* (Ft. Belvoir: Defense Technical Information Center, 1982).
13. A slow moving boat can, for example, be considered to have negligible speed with respect to a patrol aircraft and thus 'stationary' for practical purposes.
14. Strictly speaking $A = WVT + (\Pi W^2)/4$. But the latter term may be negligible if WVT is sufficiently large. This would often be the case with the velocity being high for an ariel scout even if the endurance is limited and the 'Time on Task' being large for a surface vessel.
15. The converse $(a_2/A_2) = (a_1/A_1)$ can separately be examined, with similar results.
16. Random search with locomotion permitted uses the $PoD = 1 - e^{-(a/A)}$.

CHAPTER 8

Application of Game Theory to Combat Maritime Terrorism

The threat of terrorism to maritime security is not a new phenomenon. 'Tiangong Kaiwu'[1] a Chinese treatise, dating back to 1637 AD describes naval mines fitted with a rip cord operated wheel-lock flint mechanism which could be pulled by hidden ambushers located on the nearby shore.[2] The technology could be used equally by the navies and the saboteurs.

Even concerns regarding the possible magnitude of attacks have been historically well established. As far back as 1939, Albert Einstein in a letter to US President Franklin Roosevelt wrote that "Uranium could also lead to the construction of bombs. A single bomb of this type, carried by boat and exploded in a port, might very well destroy the whole port together with some of the surrounding territory."[3]

In the Indian context, the 26/11 attack on Mumbai probably brought the issue to the fore but it must be acknowledged that numerous inputs about maritime terrorism activities by the LTTE[4] had been available for many years prior to that. As far a terrorist strike against India using the sea route are concerned, intelligence agencies had reported prior 2008, that nearly 500 LeT terrorists were being trained in coastal camps near Karachi. The six-month long maritime terror training reportedly started in October 2006. The Ministry of Defence made the following statement in a written reply to the Lok Sabha[5] on March 08, 2007[6]:

"...The intelligence agencies have warned about the possibility of terrorists trying to infiltrate through the sea route or trying to target our offshore installations. There are reports about terrorists of various tanzeems being imparted training and likelihood of their infiltration through sea routes...".

Notwithstanding the reports, the maritime domain was till recently not considered an 'attractive' target for the terrorist groups. This was despite sporadic reports on maritime terrorist training camps, the widespread acknowledgement of the seas being largely unregulated and even numerous incidents of ships having been taken over or attacked by terrorists.[7,8] An article titled 'Maritime Terrorism: India's Achilles' heel?' published by the National Maritime Foundation rues that "The issue of maritime terrorism is even today waiting to receive the attention it deserves. The famous RAND study that reported maritime terrorism as merely two percent of all international terrorist incidents worldwide is often quoted at conferences and meetings on maritime security. This figure has had many believing that maritime terrorism is a vague and distant possibility".[9]

To begin with we need to understand what maritime terrorism is, why the maritime domain was traditionally not the preferred target of attacks and also the reasons why this trend seems to be changing in the recent past.

Maritime terrorism was traditionally defined as the "use of violence at sea, to a ship or fixed platform for political ends, including any use of violence for the purpose of putting the public or any section of the public in fear".[10] The definition has undergone major changes in the recent past and The Council for Security Cooperation in the Asia Pacific (CSCAP) Working Group now defines maritime terrorism as "...the undertaking of terrorist acts and activities within the maritime environment, using or against vessels or fixed platforms at sea or in port, or against any one of their passengers or personnel, against coastal facilities or settlements, including tourist resorts, port areas and port towns or cities". While the second definition is quite exhaustive and covers the type of attacks that happened in Mumbai on 26/11, there is perhaps a need to further enhance the scope of the definition to include "any act of terror that emanates from the sea route, whether its eventual intended target is on land or within the maritime environment".[11] Use of the sea purely for logistics support of terrorist organisations, as was undertaken by the LTTE, could also be included in this enhanced scope. Even though the acts of terror themselves do not take place at sea, the use of the sea to make them possible implies that solutions can also be found at sea.

The reasons for the terrorists not taking to the sea, despite it obviously being an attractive target can be studied using the framework developed by the RAND Corporation in their study.[12] The study acknowledges that "The maritime environment possesses some unique characteristics that, in principle, could make it attractive to terrorist operations, including the extraterritoriality of the high seas and poor or inconsistent security measures that apply in coastal areas and facilities in many parts of the world". It further states that "terrorism risk does not exist without the existence of threat, the presence of vulnerability, and the potential for consequences. Terrorism threats exist if a group or individual has both the capability and intent to attack a target" and "terrorist attacks can result in human, economic, and intangible consequences. However, risks only exist if an attack can cause a change to the state of a target that results in negative consequences—in other words, if the targets are vulnerable to attack (Willis et al., 2005; Haimes, 2006)". Whilst the twin factors of 'vulnerability' and 'potential for consequences' were always present, the absence of the critical third factor of this troika, namely 'threat' was the primary reason the maritime domain was safe. Most terrorist organizations being land based had neither the capability nor indeed the intent to take to the blue waters. The 'land lubber' terrorists lacked the specialist capabilities required to mount and sustain operations from sea or the budgets required for acquiring such capability. Secondly attacks at sea had a much lower visibility and consequently lesser effectiveness than attacks on land which could be better observed and therefore well publicized. While we may accept the RAND model to study maritime terrorism, some of the findings of the study need to be debated—especially where it is claimed "many perceptions of maritime terrorism risks do not align with the reality of threats and vulnerabilities". For example, the finding that "there is little evidence that terrorists and piracy syndicates are collaborating..." may, by itself, be true but the conclusion that "maritime terrorism policy should not be motivated by these perceived threats" may be pre-mature.

Capability and intent, it must be realized, are both tangibles that can be developed given favourable circumstances. Recent developments have unfortunately provided the circumstances that could be the right motivation to develop the missing third leg of the triad. Possibly the most major cause for the need to move to the sea is increased focus on securing land-based and aviation assets. From the terrorist's perspective, the path of least resistance is thus slowly shifting seawards as the land borders are made safer. Matthew Daley, a former US deputy assistant secretary of state, summed it well when he stated[13] that, "as both the physical and political space in which they [the terrorists] find sanctuary shrinks, as the noose tightens," there is increasing likelihood of them turning "to the most unregulated of spaces—the sea."

Even then, despite the efficacy of maritime terrorism (from the terrorists' point of view) having been well established, it is difficult for the recruiters to get operatives willing to operate at sea. If we take the case of terrorism spawned by religious fundamentalism, the propaganda being undertaken to motivate cadres is apparent. The fact that 'jihadi' literature on the Internet is full of examples that seek to motivate personnel specifically for 'jihad at sea' substantiates this. Ibn Majah book of hadits known as Sunan Ibn Majah[14] includes a specific chapter on "Hadits of the Prophet on Jihad at Sea". Some of the 'motivational' propaganda includes equating one battle at sea with ten battles on the land and equating seasickness with a "fighter wounded in the path of Allah". For a terrorist brainwashed on religious ideology the promise of one martyr at sea equalling many on the land could be a big motivator, especially if he believes that "...Allah delegates the tasks of taking life to the angels of death except the life of martyrs at sea. He took it Himself. Allah forgives the sins of the martyrs on land except his debt but for martyrs at sea, forgiveness of all sins including his debt."[15]

If it is agreed that lack of maritime skills is one of the factors limiting the threat, it must be flagged that the potential for cooperation between pirates and terrorists exists. Even though there is no known proof of direct links, some analysts have sought to link the two,[16] which may not incorrect if we look below the surface. Specifically with regard to the Indian Ocean Region, some concerns have been raised about the potential links of piracy to acts of terrorism.[17] One of the factors for the connection being drawn is that two of the biggest maritime threats are being juxtaposed in geographical terms; it may only be matter of time before they converge to a symbiotic relationship. With financial support for terrorism being progressively curbed through international measures, terrorist groups may increasingly take to crime to generate funds. Though this is currently largely land based crime (such as drug smuggling or fake currency), this may not continue to be the case. Similarly with anti-piracy crackdown being effective, the pirates too would be looking for alternate sources of funds. The logic that piracy is largely undertaken for commercial gains, not political or ideological ones, does not automatically prevent a pirate from carrying out a terrorist act if someone pays them well enough—provided they have a reasonable chance of getting away. Both the scenarios are plausible, though both are contrary (either the terrorists look to piracy for funding or they themselves can fund pirates for terrorism). Piracy and suicide attacks may be contrary but piracy and terrorism are not. When the marine police was being set up in India in the aftermath of 26/11 and it was realized that conventional policemen lack 'sea legs', the obvious suggestions were engaging retired naval personnel or induction of personnel from coastal traditional fishing communities into the force. Would such an obvious solution not be apparent to the terrorists? Conventional wisdom dictates an eventual piracy-terrorism nexus that could well bridge the gap between the intentions of the terrorist groups and their maritime capabilities.

Once the decision to take the marine route is taken, the sea also provides many advantages that are immediately apparent. Since the law enforcement agencies would always have limited resources, the expanse of the sea and large numbers of potential targets make maritime terrorism attractive. The unregulated sea, largely opaque and amorphous, is also a far safer option from the terrorists' perspective. Seasickness, whilst being equated to being wounded by a bullet, would certainly be the preferred option between the two being the lesser of the two evils!

In a globalised world, where the world economy is largely sea based, attacks at sea need no longer be less visible. A high profile maritime asset with great economic value or the threat of economic chaos that could result from targeting selected choke points (say by sinking of a ship in the narrow entrance to a major port) would generate the kind of attention that terrorists seek. An article 'Marine Terrorism: A Strategic Necessity', that appeared on Ekhlass[18] website opined that "It has become very crucial to develop the battlefield to reach the sea". The same was widely reported in many radical forums. For example the website armiesofliberation.com noted "We hope from our brothers in Yemen to get us back to the old days of attacking USS Cole and French Limburg as such operations influence the people nowadays".[19]

The last factor that can drive the maritime component of terrorism higher is the spread of technology. Availability of the GPS (Global Positioning System) makes navigation at sea simpler and less specialised, similarly satellite phones ease the command and control problems, AIS (Automatic Identification System) makes target identification and tracking easier etc. More importantly the Internet itself has become a powerful knowledge tool for the terrorists. Information that was hard to come by is now easily available, as are destructive ideas. The tactics of one organization are closely studied by another in an increasingly globalised world leading to a cross pollination of ideas. This enables the groups to replicate not only attacks but also adapt land capabilities in the maritime environment without much difficulty. This "copycat effect"[20] was demonstrated post the USS Cole attack on 12 October 2000 which was followed, less than 20 days later, by LTTE destroying a fast personnel carrier on 23 October 2000 and another similar, though failed, attack by Hamas suicide boat on an Israeli naval craft on 07 November 2000.[21] As one analyst has noted "in today's information age, the terrorists' transnational network enables them to share the operational tactics of contemporary, proven attack methods, and also encourages other terrorist organizations to adopt such methods. Thus, the possibility of small boat terror attacks, similar to the 2008 Mumbai terror attacks, in the future and against other countries, especially those combating terrorism, cannot be ignored".

All these factors are likely to combine to make the threat of maritime terrorism greater unless effective steps are taken to thwart the same. It is also important that while only 2% of terrorist attacks worldwide may have been maritime in nature, regional statistics may show a completely different picture. For example, three out of the four major terrorist attacks in Israel in the 1970s were maritime infiltration attacks.[22] This number went down substantially after concerted steps were taken to control these by a combination of defensive as well as offensive actions. Israel, as the foremost target of maritime terrorism, developed in the 1970s what today is one of the best coastal defences worldwide.[23] Efficacy of their coastal security has been well established over the past decades with Israel detecting (according to a senior Israeli Navy officer) more than 80 maritime terror plots.[24] On May 30, 1990 Israeli security forces foiled an attempt by PLF terrorists to infiltrate Israel from the sea.[25] The terrorist operation code named Al-Quds involved 16 terrorists. According to PLF leader Abu Abbas, the preparation for this operation had taken nearly three years and cost about three million dollars. Following the failed plot, the PLO announced through Kaled al-Hassan on June 4, 1990, that the PLF operation was utilizing the wrong technique "because everyone knows that Israel has stated that its radar lab can scan as far as Malta. From the technical point of view it was an act of madness".[26] Adversary impression about security forces capability is possibly more decisive in the battle against terrorism than the capability itself. The terrorists are unlikely to attempt another "act of madness" if they believe that the chances of success are nearly nil.

Some distinct characteristics of maritime terror must next be acknowledged. The most prominent of these being the need to operate from a safe haven ashore which may be within the country (as in the case of erstwhile LTTE controlled Northern areas within Sri Lanka) or more likely in foreign shores. This distinction is critical as the two players in the terrorist game will likely be independent states. In the words of a Sri Lankan Navy Admiral addressing the 19th International Sea Power Symposium in Rhode Island, U.S., "A terrorist organization with maritime capability needs to operate from foreign soil or a safe base; the prevention of the use of foreign soil for all illegal activity particularly on remote islands and isolated coastal stretches needs to be addressed."[27]

In the Indian context the following merits attention:

- It is likely that the threat would emanate from across the border if we consider the most likely types of attack including infiltration of terrorists and/or explosives into the country. Use of the sea route to move terrorists/explosives from say Gujarat to Mumbai makes little sense given the complexities discussed previously. Use of the land route would be the preferred option being much easier. The same is not true for cross border movements where the sea route may have advantages.

- Use of a mother ship or containerized cargo for the purpose is feasible though the chances of success would be adversely affected by the many security measures put into place post 26/11. Small boat terrorism may continue to be the biggest threat.
- Given the complexity of a transnational operation, the chances of success increase with state backing (both in terms of resources as well as training). Small non-state actors may be incapable of mounting large operations by themselves.
- A USS Cole type of an incident, of domestic origin, is unlikely but cannot be ruled out. The quantum of explosives required for such an attack implies that the explosives would have to smuggled in into the country in the first place—thereby making such an attack also related to the three identified previously.

To study the problem from a Game Theory perspective, a similar study undertaken by Captain Jaswinder Singh can be used as the base. The study 'Countering Small Boat Terrorism in Territorial Sea'[28] utilizes Game Theory to analyze various counterterrorism strategies, and infers how "security forces could tilt the game of small boat terror attacks to their advantage" (figure 8.1). Counter-terrorism strategies have been divided into offensive and protective and the options available for both the terror origin state and the terror victim state modelled as a simple 2×2 game in the normal form (figure 8.2).

The options available for the terror victim state terror origin state include options (A) Offensive and (B) Protective. The terror origin state similarly has to options (C) Take no action against the terrorists and (D) Take action against the terrorists.

When the four possible results are analysed after giving each player a payoff depending on the options being graded as best, next best, least best and worst (respective payoffs being 4,3,2, and 1), the game without the players communicating would be as depicted in figure 8.2. Each player has a pure dominant strategy. The terror origin state's pure dominant strategy is not to take any action against the terrorists and the terror victim state's pure dominant strategy is to remain protective. The Nash Equilibrium of this game is (BC), (3, 4). This implies that terror origin state will take no action against the terrorists and the terror victim state will continue to remain protective.

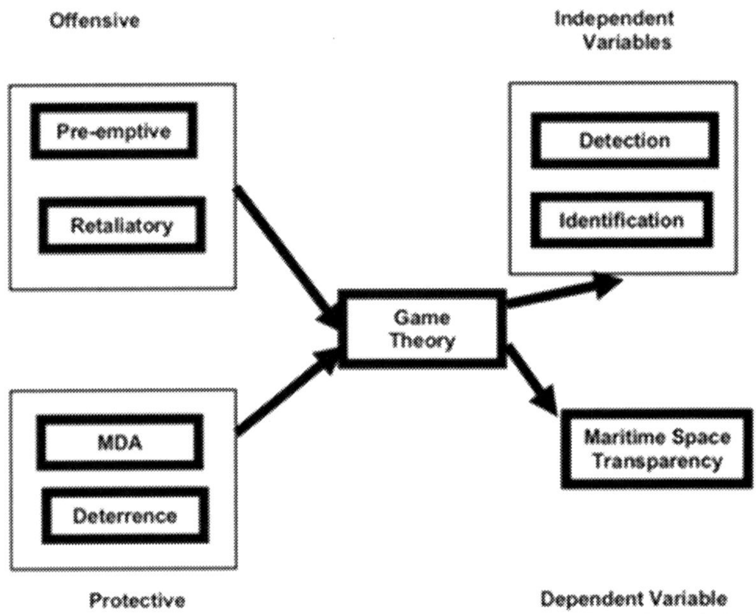

Figure 8.1

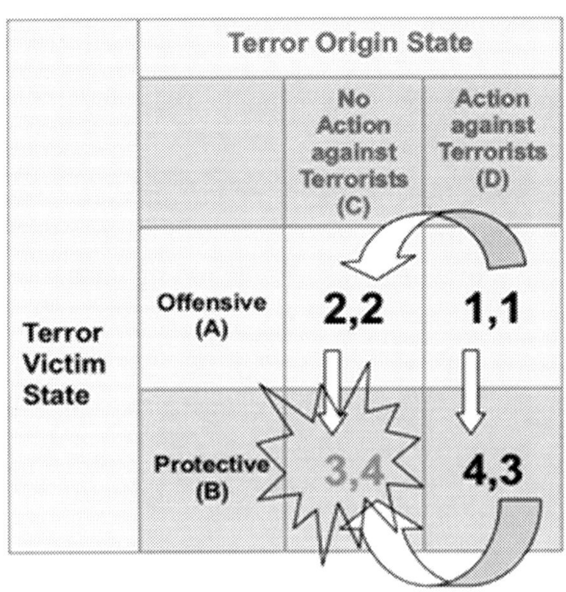

Figure 8.2

Some of the interesting findings of the study are that the basic solution 'BC' does not alter even if the players are allowed to communicate and either player given the first mover status. The two questions "Can Terror Victim State Threaten Terror Origin State?" and "Can Terror Victim State Promise Terror Origin State?" have also been analysed with the conclusion that "Threat exists, as the threatened outcome hurts both the players" and that "Terror victim state can't promise terror origin state, as the promised outcome would be worst for both the players". Of course the terror origin state threatening or promising the terror victim state may be meaningless as the present outcome being the best from its perspective, the terror origin state would not like to threaten or promise.

The study further evaluates the game with Interval Scaling from 1-10 with payoffs as tabulated below:

Terror Victim State	*Terror Origin State*
10 – Best – Terror origin state takes action against the terrorists; terror victim state remains protective	10 – Best – Terror origin state takes no action against the terrorists; terror victim state remains protective
5 – Next Best – Terror origin state takes no action against the terrorists; terror victim state remains protective (since the terrorists could infiltrate the terror victim state's MDA, low absolute value of the protective stance has been accorded)	9 – Next Best – Terror origin state takes action against the terrorists; terror victim state remains protective (considering that terror origin state would have to put in some effort to take action against its citizens)
4 – Least Best – Terror origin state takes no action against the terrorists; terror victim state goes offensive (considering that terror victim state would be in an advantageous position compared to the terror origin state)	2 – Least Best – Terror origin state takes no action against the terrorists; terror victim state goes offensive (considering that terror origin state would be in a disadvantageous position compared to terror victim state)
1 – Worst – Terror origin state takes action against the terrorists: terror victim state goes offensive	1 – Worst – Terror origin state takes action against the terrorists: terror victim state goes offensive

The results of the interval scaling game are depicted in figure 8.3, and the final outcome, it may be noted, remains unchanged.

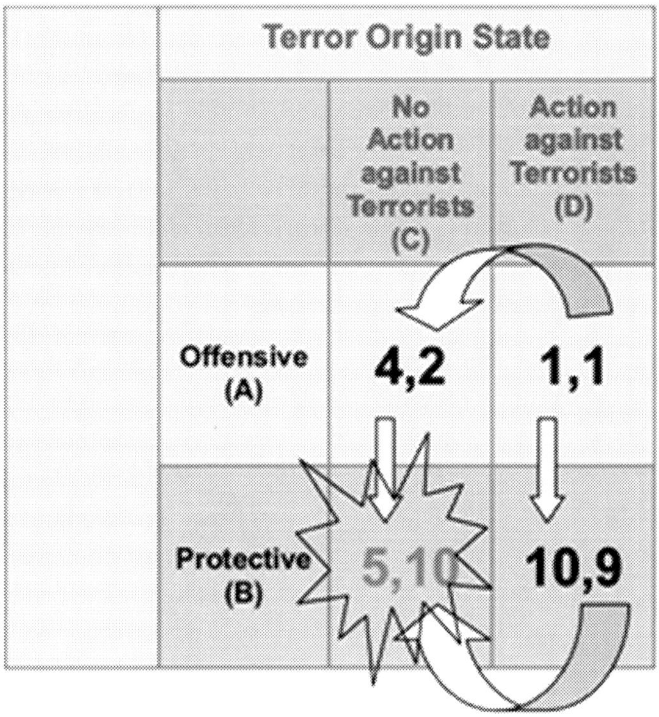

Figure 8.3

This picture, along with the analyzed strategic moves, conveys that Terror origin state's pure dominant strategy is not to take action against the terrorists (absolute payoff of 10) and the terror victim state's pure dominant strategy is to remain protective (absolute payoff of 5). Additionally it is brought out that the victim state could coerce terror origin state to take action against the terrorists, as this game has a threat and moreover since there is only slight reduction in terror origin state's payoff (i.e., from an absolute value of 10 to 9). From this it is concluded that the terror victim state should increase its BC Strategy's (protective strategy even if terror origin state takes no action against the terrorists) payoff, which is presently 5, by improving its MDA in territorial sea or in other words by strengthening its defensive posture. Both these findings merit further deliberation.

Even when the findings are analysed from the security level perspective, as opposed to a pure economics standpoint, where both the players are maximizing their respective payoffs, the picture does not substantially change. The security levels are the minimum assured payoffs for each player when the other player starts minimizing the opponent. This gives the terror victim state a security level (minimum assured payoff) of 5, and the prudential strategy is to remain protective (figure 8.4a). When the terror origin state is maximizing its payoffs, whereas terror victim state is minimizing them, terror origin state has a security level (minimum assured payoff) of 2, and the prudential strategy is not to take action against terrorists (figure 8.4b).

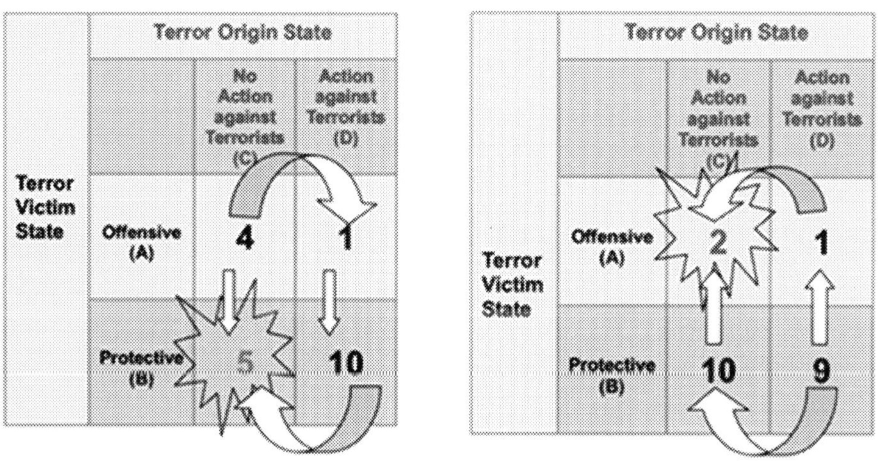

Figure 8.4 (a), 8.4 (b)

None of the equilibriums considered till now is Pareto optimal. The Nash Point, a Pareto optimal solution of this game is (10, 9) implying solution BD where the terror origin state should take action against the terrorists and terror victim state should continue the protective strategy (figure 8.5).The best that the victim state can do by enhancing its MDA is to bring the security level up to (9,2). This may make the task of the origin state more difficult but does not substantially alter the security level per se.

The data generated by the study, taken at face value and accepting the basic assumptions to be correct implies the following (if we try and reach conclusions independently of what has been reached by the study):

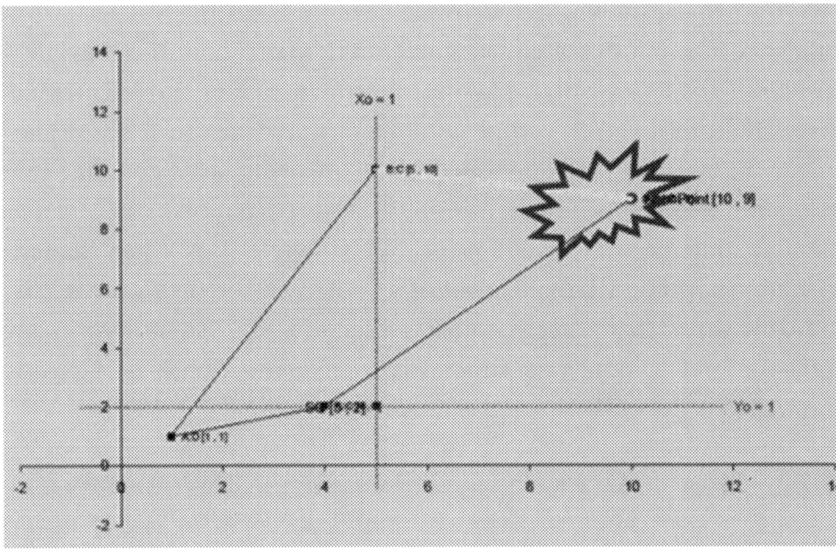

Figure 8.5

(a) If the primary assumption in deciding the payoffs, namely that aggressive action is less desirable that protective measures, is valid then the best solution for the victim state shall continue to be to remain defensive.

(b) No matter what the victim state does, to strengthen its defences, as long as offensive action is ruled out the origin state will continue to support terrorism.

(c) The study acknowledges that potential for threats and coercion exists but these are not taken as separate options for the victim state.

The results of any game would lead to the conclusion that enhancing MDA is the only option till we can come up with solutions that increase the 'cost' of supporting terrorism for the country supporting terrorism. A purely protective strategy implies that it is only a matter of time before the next vulnerability is discovered and exploited. Even if the same is strengthened subsequently, new vulnerabilities will subsequently be emerging leading to a reactive cycle where 'Every vulnerability will eventually be found and every vulnerability will eventually be plugged'.

Interestingly another study, focused not on maritime terrorism but on the application of Game Theory for countering the Lashkar-e-Taiba[29] (LeT) reached the conclusion that the Pakistani civilian government could remain passive and yet the LeT could be coerced into eliminating its armed wing. The complex game included five players (the LeT, India, United States, Pakistan Military and Pakistan Government). Since each of the players could choose multiple options, a total of 1916 Nash Equilibria were identified out of which the six pure equilibriums that resulted in a high probability of LeT disarming are tabulated below. The said study inferred that "joint application of pressure by India and the US on Pakistan (through covert actions and/or coercive diplomacy) is key to incentivizing responsible behaviour not only from LeT, but also the Pakistani military".[30]

Eq.	US	India	Pak-Military	Pak-Government	LeT
E_1	Cut Support	Covert Ops	Cease Support	None	Eliminate A.W.
E_2	Covert Ops	Coercive Diplomacy	Cease Support	None	Eliminate A.W.
E_3	Covert Ops	Coercive Diplomacy	Cease Support	Prosecute LeT	Eliminate A.W.
E_4	Covert Ops	Coercive Diplomacy	Crackdown LeT	None	Eliminate A.W.
E_5	Covert Ops	Covert Ops	Cease Support	None	Eliminate A.W.
E_6	Covert Ops	Covert Ops	None	Prosecute LeT	Eliminate A.W.

Since the support from Pakistan military would not cease till some external factors forced it to do so, the question becomes what can be done to achieve this end.

Whilst exploring the possible options we do not have to look far. Though Game Theory is a western concept, the Indian *Chanakyaneeti*[31] has several tenets that can be used for providing the range of options for settling conflicts. The ancient doctrine explains the four pronged approach of 'sama-dam-dand-bhed' (roughly translated as conciliation-economic incentives-punishment-clandestine divisive means)[32] towards the adversary. The four methods can be used singly, or in multiple combinations. Choosing one, two, three or all four of these methods for dealing with any conflict provides fifteen combinations of tactics for dealing with the conflict. The *neeti* does not restrict the tenets but indeed is a broad based explanation having multiple interpretations of the four terms. The term Dam for example may be taken to mean offering a gift or a bribe, it can equally refer to imposing economic hardships as is done by the UN to coax recalcitrant regimes by imposing sanctions. India can certainly impose a set of calibrated economic costs on Pakistan, similarly coercive diplomacy and covert or overt actions are possibilities. For the threat to be valid the other player must be equally aware of the options, the real possibility of their being invoked and also the possible outcomes. A classified plan to ensure punitive retaliation may not be effective as a threat until the adversary is aware of it and believes it to be possible.

A normal form two player game with four options for India roughly corresponding to 'sama-dam-dand-bhed' can be constructed with a view to explore the outcomes if punitive measures are to be imposed. Coercive diplomacy, economic sanctions, overt punitive action including the oft discussed 'surgical strikes' and covert action may be taken as the four options. The payoffs can be accorded on a scale of 1 to 5 as under:

1 Worst
2 Bad
3 Neutral, with losses and gains being approximately matched
4 Good
5 Best

It is difficult to assign payoffs in such a game without an element of subjectively. However since the measures to be taken by India are by definition 'punitive' it is assumed that the payoffs for the other player vis a vis 'business as usual' would be lower. Use of all the four options can thus be modelled to generate a calibrated response.

Use of Game Theory for countering state sponsored terror can be undertaken by considering the sponsor state as one of the players. This needs to be undertaken for answering larger foreign policy issues as has also been propagated by other commentators. Chief Economic Advisor Kaushik Basu in his talk at the IDSA in November 2010, for example, emphasized upon the need for Indian think tanks to use Game Theory in modelling foreign policy behaviour.[33]

NOTES

1. Roughly translated as 'The Exploitation of the Works of Nature'.
2. Joseph Needham, *Science and Civilisation in China Volume 5*, Part 7, p. 205.
3. http://strategyonblog.blogspot.in/2012/04/maritime-terrorism-introduction-uranium.html accessed 13 February 2013.
4. LTTE or the Liberation Tigers of Tamil Elam were a Sri Lanka based separatist group

with a dedicated sea wing called the Sea Tigers.

5. One of the two Houses of the Indian Parliament.
6. Press Information Bureau Government of India accessed 13 February 2013 at http://pib.nic.in/release/rel_print_page1.asp?relid=25642
7. Pyeatt, Menefee S., "Piracy, Terrorism, and the Insurgent Passenger: A Historical and Legal Perspective," *Maritime Terrorism and International Law, (1990): 43-68*
8. Acharya, Arabinda and Withana, Nadeeka Prashadani, *Groups with Maritime Terrorist Capabilities in the Indian Ocean Region*. Raghavan, V.R. and Prabhakar, Lawrence, *Maritime Security in the Indian Ocean Region Critical Issues in Debate*, Centre for Strategic Studies (New Delhi, Tata McGraw Hill Publishers, 2008).
9. Riddhi Shah, *Maritime Terrorism: India's Achilles' Heel?*, accessed 23 February 2014 at http://www.maritimeindia.org/article/maritime-terrorism-india%E2%80%99s-achilles%E2%80%99-heel
10. CSCAP memorandum number 5, Cooperation for Law and Order at Sea, February 2001.
11. Shashikumar, V.K., "Gaps in Maritime Security," *Indian Defence Review*, Vol 24.1 Jan-Mar 2009 accessed at http://idr.qburst.com/interviews/gaps-in-maritime-security-i/ on 13 February 2013.
12. Greenberg, Michael D., Chalk, Peter, *et al.*, *Maritime Terrorism: Risk and Liability.* (Santa Monica, CA: RAND Centre for Terrorism Risk Management Policy, 2006).
13. http://www.hurriyetdailynews.com/default.aspx?pageid=438&n=asia-eyes-sea-security-after-terrorist-threats-2004-04-23 accessed on 13 February 2013.
14. http://archive.org/stream/SunanIbnMajah5Vol.Set/Sunan-Ibn-Majah-Volume-4-English_djvu.txt accessed on 13 February 2013.
15. Ibid.
16. Burnett, John S., *Dangerous Waters: Modern Piracy and Terror on the High Seas* (New York, N.Y: Dutton, 2002).
17. Donald L Berlin, *The Evolving Trends in Maritime Security, Maritime Security in the Indian Ocean region: Critical Issues in Debate*, Edited by V.R. Raghavan, W. Lawrence Prabhakar, Centre for Security Analysis (New Delhi, Tata McGraw-Hill)
18. The al-Ekhlass website is an Arabic language jihadist website which also has an English language section.
19. http://armiesofliberation.com/archives/2008/09/21/al-qaeda-in-yemen-timeline/ accessed 01 Sep 2013.
20. http://strategyonblog.blogspot.in/2012/04/maritime-terrorism-introduction-uranium.html accessed 01 Sep 2013.
21. The suicide boat exploded prematurely and could not impact the naval vessel.
22. Akiva J. Lorenz, "The Threat of Maritime Terrorism to Israel," *International Institute for Counterterrorism*, September 24, 2007, accessed on 07 Feb 2013 at http://www.ict.org.il/Articles/tabid/66/Articlsid/251/currentpage/6/Default.aspx
23. Gunaratna, Rohan, *The Threat to the Maritime Domain: How Real is the Terrorist Threat?*, http://www.nwc.navy.mil/nsdm/Rugerpapers.htm accessed 30 August 13.
24. Zadeh, Ali Nouri, *Iranian officer: Hezbollah has a Commando Naval Unit*, July 29, 2006, http://www.asharqalawsat.com/english/news.asp?section=1&id=5801 accessed

30 August 13.

25. Akiva J. Lorenz, *The Threat of Maritime Terrorism to Israel.*
26. Ibid.
27. Samarasinghe, Thisra, "Sri Lankan Navy's role in eradicating international maritime terrorism," *Sri Lanka Guardian*, October 25, 2009 accessed on 30 August 2013 at http://www.srilankaguardian.org/2009/10/sri-lankan-navys-role-in-eradicating.html
28. Singh, Jaswinder, *Countering Small Boat Terrorism in Territorial Sea* (Naval Postgraduate School, Monterey, CA, United States (2010).
29. Dickerson, John P., Mannes, Aaron, Subrahmanian, V.S., *Dealing with Lashkar-e-Taiba: A Multi-Player Game-Theoretic Perspective.*
30. Ibid.
31. Roughly translated as the treatise of Chanakya.
32. Chanakya in the Arthashastra and as translated by LN Rangarajan, Penguin Classics
33. Arvind Gupta, *Game-theoretic Understanding of India-Pakistan Relations*, accessed http://www.idsa.in/idsacomments/GametheoreticUnderstandingofIndiaPakistanRelations_agupta_160811

CHAPTER 9

Application of Fuzzy Game Theory Concepts to Triadic Deterrence: Tackling State Sponsored Terrorism

"As complexity rises, precise statements lose meaning and meaningful statements lose precision"

—Lotfi Zadeh

India has been beset with the problem of tackling state sponsored terrorism emanating from across the border for the last few decades. Some commentators have opined that 'Non-state', and possibly state-sponsored actors have proliferated due to what nuclear theorists call the stability-instability paradox. Nuclear deterrence has, as per these analysts, created the space for sub-conventional conflict. This implies that while the presence of nuclear weapons has decreased the likelihood of direct war, it may by itself provide more room for inciting India without fear of retaliation.[1]

The Indian reaction to this escalation in the sub-conventional domain has largely been restrained. Sunil Dasgupta and Stephen P. Cohen commented that "one of the most remarkable attributes of India as an independent state has been its reticence to use force as an instrument of policy".[2] There is little doubt that this 'reticence' to use force has helped the country being viewed as a responsible nuclear power in the eyes of the world community and has yielded many benefits. However some commentators have questioned what they consider an 'idealistic' and 'utopian' policy that "suggests that India's influence is ideational and moral rather than material. The problem is that while ideas matter, it is less important than material power and usually its servant. Morality, "the power of example", is even more problematic because it is inconsequential in international politics".[3]

Thomas Schelling, while advocating that the concept that military strategy can no longer be defined as the science of military victory, argued that military strategy was equally the art of coercion, of intimidation and deterrence. Deterrence has been defined as "influencing the choices that another party will make, and doing it by influencing his expectations of how we will behave".[4] The basis of effective deterrence, whether general or immediate, is that the 'power to hurt' is really successful only when it does not have to be used. Such power, held in reserve, should be sufficient for persuasion of the adversary to alter potential behaviour.[5] According to another definition, deterrence is "the prevention from action by fear of the consequences. Deterrence is a state of mind brought about by the existence of a credible threat of unacceptable counteraction".[6] Deterrence requires four necessary conditions—a clearly defined unacceptable behavior; communication of commitment to punish violations; the means to inflict punishment; and the plausible resolve to carry out the punishment if deterrence fails.[7] Successful deterrence therefore goes beyond 'capability' alone and is equally impacted by 'credibility'.

Most of the early work in deterrence theory was focused on the superpower relationship against a nuclear backdrop. This type of deterrence was clearly mutual, and was well intended to be, as is obvious from the use terms such as Mutually Assured Destruction (MAD). Deterrence theories, such as Brody's (1959)[8] and Kahn's (1965)[9] which evolved during the cold war were focused on these aspects.

In the post-Cold War years, concepts of deterrence theory were increasingly applied to conventional warfare, even between unequal foes. Fearon's Rationalist Explanations for War, for example, used concepts popularized by deterrence theory to study the decision to go to war as a bargaining model. As per this, the ex ante (pre-war) and ex post (after war) costs and how they apply are weighed by the national leadership before plunging the country into war.[10] Obviously a country that is weaker, and therefore less likely to succeed, is also less likely to initiate a war. After

9/11 terrorist attacks, many divergent streams of work on deterrence theory gained momentum. With regard to deterrence and its application in countering terrorism, A RAND study[11] of 2002 stated that "after reviewing the issues, we concluded that Cold War deterrence theory was not, in fact, a very good model for our purposes". While the existing body of work was still relevant, it needed to be enhanced and built upon to reflect the changed realities. Fuzzy Deterrence,[12] Deterrence 2.0[13] or The Fourth Wave[14] in Deterrence Research were some of the new 'buzzwords'. These theories recognized that scaled response and calibrated escalation were also very much a part of deterrence and coercion. To tackle the new enemy, namely non-state actors, the question of whether deterrence could be tailored[15] was obvious. In a scenario in which the non-state actors, the sponsor state and the victim state are involved, the deterrence game becomes three sided. Such deterrence can nonetheless be studied as a two player game if it appears likely that the terror origin state has the ability to control the non-state actors which may be acting as the proxy agents of the state. Triadic deterrence is the situation when one state uses threats and/or punishments against another state to coerce it to prevent non-state actors from conducting attacks from its territory.[16]

Non-state actors (whether state sponsored, supported or otherwise) have certain peculiarities that may make direct application of deterrence to them a difficult proposition. Non state actors, unlike nation-states, normally seek to change the status quo,[17] and would be willing to adopt extreme measures. Their behaviour can therefore be difficult to predict. Further, they do not exercise territorial sovereignty[18] and lack clearly identifiable centers of gravity.[19] Holding territory to risk for threatening that control or targeted attacks against specific objectives may not be feasible. Triadic deterrence, which targets the sponsor state, may be an alternate solution. Such deterrence, it may be argued, can only be partially effective. A less than ideal deterrent should nonetheless be preferred to no deterrence.

When examining state sponsored terrorism and non-state actors, a distinction between the two needs to be made. Non state actors may operate within the safe haven provided by a state which is incapable of tackling them. In such a case holding the state that lacks the means to control the group(s) may be unjustified. State sponsored terrorism on the other hand assumes active involvement of the state to encourage these groups and in fact goes to the extent of using them as instruments of state policy. Three conditions must be met before such a situation arises.[20] First of all there has to be an adversarial relationship between the sponsoring and victim state or a motivation for conflict. Secondly there must be present some constraints against pursuing conflict openly, which may due to military superiority of the adversary, international pressure or other reasons. Lastly the sponsor state must enjoy a lack of accountability for their conduct. It is important to note that all three conditions are necessary and must co-exist for sponsorship to occur. It also therefore follows that removing one or more of the conditions can cause the sponsorship to cease.[21] Removing the motivation for conflict would entail improving bilateral ties between the two countries and whilst being the ideal solution may not be feasible in the short term and could even be idealistic or Utopian in some cases. The second basic requirement of constraints against open conflict is important. In most cases such constraints would be due to military asymmetry. Terrorism, guerrilla warfare and conventional war must be seen a continuum along the entire spectrum of options that would be used by the weaker state in the face of a stronger adversary and the option selected would be based on the level of asymmetry.[22] It has also been argued that as the gap reduces, the escalation of violence to the next higher level would almost invariably follow.[23] The victim attempting to alter the second condition would therefore be self-defeating. It is however important to note that if this condition is assumed to exist, any action by the victim state that forces the sponsor state to make a choice may result in de-escalation due to the sponsor state capitulating. The only other way to alter the status quo is to increase the accountability of the sponsor state. This means not only removing the veil of anonymity (or supposed anonymity) but also sending a message that such behaviour would invite counter action.

Deterrence between states in the scenario being examined would be best studied by using the unilateral deterrence game as this allows differentiation between states in an unequal relationship.[24] One such examination using a simple unilateral deterrence game is depicted below.[25]

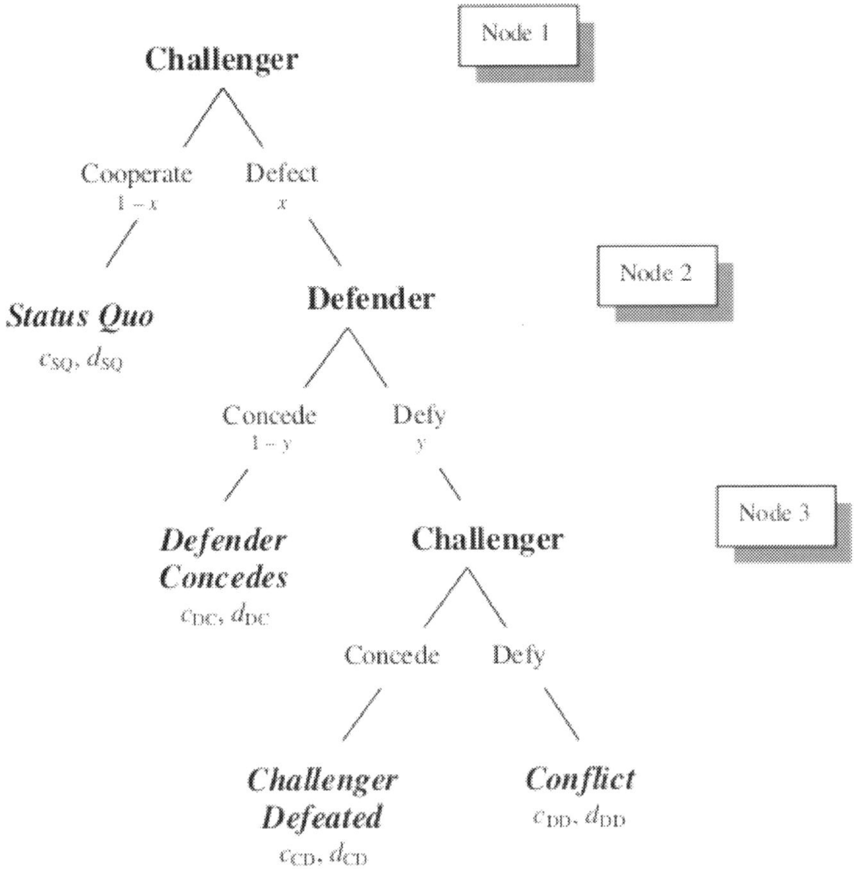

This game assumes two players. The challenger, acting first, can defect with a probability 'x' and the defender can then choose to respond by conceding or defying (probability 'y'). Based on the response, either the defender concedes (DC) and is termed a 'soft state' or defies. At node three the challenger is faced with a similar choice of whether to concede or defy.

Based on a number of factors including presence of a capable threat and the intrinsic preferences of the players ('soft' versus 'hard' state) a number of possible outcomes is presented by the study cited as is tabulated on next page.[26]

If the basic presumption that the state sponsoring terrorism is restricted in openly initiating conflict is taken to be true, the number of possibilities

Existence conditions	Restrictions on challenger	Restrictions on defender	Predicted outcome
1) $c_{SQ} > c_{DC}$	–	–	*Status Quo*
2) $c_{DD} > c_{SQ}$	–	$d_{DD} > d_{DC}$	*Conflict*
	–	$d_{DC} > d_{DD}$	*Defender Concedes*
3) $c_{DC} > c_{SQ}$ and $c_{SQ} > c_{DD}$			
Certain Deterrence Equilibrium			
$p_{Def} \geq c_t$	–	–	*Status Quo*
Separating Equilibrium			
$c_s \leq p_{Def} \leq c_t$	$c_{DD} > c_{CD}$	$d_{DD} > d_{DC}$	*Status Quo*, *Conflict*
		$d_{DC} > d_{DD}$	*Status Quo*, *Defender Concedes*
	$c_{CD} > c_{DD}$	–	*Status Quo*
Bluff Equilibrium			
$p_{Def} < c_s$ and $p_{Ch} \leq d_n$	$c_{DD} > c_{CD}$	$d_{DD} > d_{DC}$	*Status Quo*, *Conflict*
		$d_{DC} > d_{DD}$	*Status Quo*, *Defender Concedes*, *Conflict*
	$c_{CD} > c_{DD}$	$d_{DD} > d_{DC}$	*Status Quo*, *Challenger Defeated*
		$d_{DC} > d_{DD}$	*Status Quo*, *Defender Concedes*, *Challenger Defeated*
Attack Equilibrium			
$p_{Def} < c_s$ and $p_{Ch} \geq d_n$	$c_{DD} > c_{CD}$	$d_{DD} > d_{DC}$	*Status Quo*, *Conflict*
	$c_{CD} > c_{DD}$		*Status Quo*, *Challenger Defeated*
	–	$d_{DC} > d_{DD}$	*Status Quo*, *Defender Concedes*

can however be significantly reduced. Faced with a choice at node three, the state that was constrained in initiating conflict may also be assumed to be equally constrained in meeting the challenge thrown. As per this logic, the terror victim state, India in our context, should logically follow an aggressive posture. Many would argue that such a simplistic approach may be dangerous. Conflicts happen after deterrence has failed not because the states involved did not think through their actions but because situations are permitted to overtake rational decision making and factors such as nationalistic pride and public opinion start dictating policies, even if they may have disastrous consequences.

Arguing that in a dangerous world, we need to think very carefully about how military force is used,[27] Roger B. Myerson questions "What determines why an application of military force, which was intended to deter our potential adversaries, sometimes instead stimulates them to more militant reactions against us? When we feel that force is necessary, what can we do to minimize the risk of such adverse reactions against us?"[28] A successful deterrent strategy requires a balance between resolve and restraint. For this balance to be recognized and understood by our adversaries our forceful actions should be "framed by a process of communication with our potential adversaries that establishes mutually recognized limits and rules about what we will and will not do".[29] A simple two player Prisoners' Dilemma game is used in the study cited. Player 'A' represents America and player 'B' some 'other smaller country'. Each player, acting simultaneously, must choose between cooperation or aggression. Assigning the expected payoffs in such a scenario can be tricky and the values used in the study could well be debated. The dilemma that while mutual cooperation yields a better outcome than mutual defection, but it is still not the rational outcome, however, remains valid. The Nash Equilibrium (the only outcome from which each player could only do worse by unilaterally changing strategy) continues to be mutual defection because the choice to cooperate, at the individual level, is not rational from a self-interest point of view. The expected payoffs used in the study are tabulated below and the best payoff that each player could get in response to each possible action of the other player have been highlighted with an asterisk (*).

	B cooperative	B aggressive
A cooperative	0, 0	–8, 1*
A aggressive	*1,–8	*–3,–3*

A's payoff, B's payoff

The strategy[30] for player 'A' in a scenario where the game is not simultaneous, but where 'A' responds to the actions of 'B' are tabulated below:

A's strategy:	*B's action:* *B cooperative*	*B aggressive*	B cooperative	B aggressive
A is cooperative always	A cooperative	A cooperative	0, 0	–8, 1*
A does the same as B	A cooperative	A aggressive	0, 0*	*–3,–3
A does opposite of B	A aggressive	A cooperative	*1,–8	–8, 1*
A is aggressive always	A aggressive	A aggressive	*1,–8	*–3,–3*

In that particular row, the asterisk (*) after the second number in a cell is indicative of the best payoff that B can get in response to the strategy of A. Only one cooperative payoff for B emerges as the best option for B—the strategy where A 'does the same as B'. This would be the only deterrent strategy for A that encourages cooperative behaviour from B. This strategy, by itself, is however non credible as the payoffs for A are higher in two other alternatives, namely 'aggressive always' and 'opposite of B'. Unless A can somehow be constrained to follow this strategy, the game has only one equilibrium.

A nation that is weaker (geographically smaller, economically or militarily disadvantaged) may perceive, justifiably or not, a threat for its existence from the stronger adversary. In the nuclear realm the 'no first use policy' may be seen as a self-imposed restraint to enable cooperative behaviour. Massive retaliation, in case of defection is used to make this credible. There have also been other instances of such self-imposed restraints being gainfully utilized by India even in the conventional domain. For example, during the Kargil conflict, the decision to use airpower coupled with the declaration of respecting the Line of Control (LoC) enabled India to up the ante militarily without risk of nuclear escalation. The nuclear bluff by Pakistan in that scenario was thereby not credible.

Tailored deterrence comprising 'scaled response' and 'calibrated escalation' can, therefore, offer an alternative to the simple game of Prisoner's Dilemma discussed above. Rather than the binary 'yes' or 'no' answers, deterrence may be considered as a continuum with incremental changes in the level of action by the state seeking to deter an adversary. Deterrence by dissuasion,[31] deterrence by denial[32] and finally deterrence by threat[33] are all valid forms of shaping adversary behaviour. Even the most extreme case of deterrence by threat can range from the low (targeted strike) to the high (invasion) end of conflict.[34] Flexibility, adaptability, and ambiguity which are the basic foundations of 'fuzzy logic' can thus lay the foundation for triadic deterrence theory.

Fuzzy logic is a form of many-valued logic that deals with reasoning that is approximate rather than fixed and exact. The British philosopher and logician Bertrand Russell published (in 1923) an article aptly titled—*Vagueness*. Russell asserted that language is not precise, but vague and vague to a matter of degree.[35] Fuzzy logic, as an extension of Russell's concept of vagueness, was incorporated as a mathematical concept in the fuzzy set theory proposed by Lotfi A. Zadeh in 1965. This provided a method to move beyond classical logic's strict binary 'true' or 'false' depiction of conditions. For example water may be either 'hot' or 'cold' in binary terms, but could be described by a number of degrees of 'hotness' from warm, luke warm, tipid, slightly hot etc. Cold could similarly be 'somewhat' cold or 'very' cold or the even more vague 'nippy'. Similarly, even in military terms the question does not have to be whether to attack or not, but if military victory is not taken as the sole criteria, also 'how much' to push the adversary and by what means. The action need not follow the if-then-else paradigm. To take another example, a simple temperature regulator that uses a fan might use the rule set[36]

IF temperature IS very cold THEN stop fan
IF temperature IS cold THEN turn down fan
IF temperature IS normal THEN maintain level
IF temperature IS hot THEN speed up fan

Obviously, in fuzzy terms, the temperature may simultaneously be 'cold' and 'normal' to different degrees. Therefore all of the rules (which at least theoretically may be infinite) must be individually evaluated and there is no room for the 'ELSE' operator.[37]

When applied to triadic deterrence, fuzzy logic permits an element of uncertainty whilst also accommodating the range of options available to both the players. Taking, once again, the simple Prisoner's Dilemma game (Myerson, 2006), the options available for the terror victim state are either to cooperate or be aggressive. It is apparent that the binary payoffs, in pure mathematical terms, of –8 and –3 respectively can be observed if an aggressive terror sponsor is encountered. This implies that up to a certain point the payoff is negative and no aggressive action is taken by the victim and further that this shifts to –3 at that point. In actual terms however the level of aggression by the victim state can be calibrated, so as to obtain a curve between the two extremes as the aggression increases from left to right in the graph below. A reduction in the positive payoff for the sponsor is correspondingly depicted by the upper curve. With the payoffs being equal in the Nash Equilibrium, both curves coincide at (–3,–3).

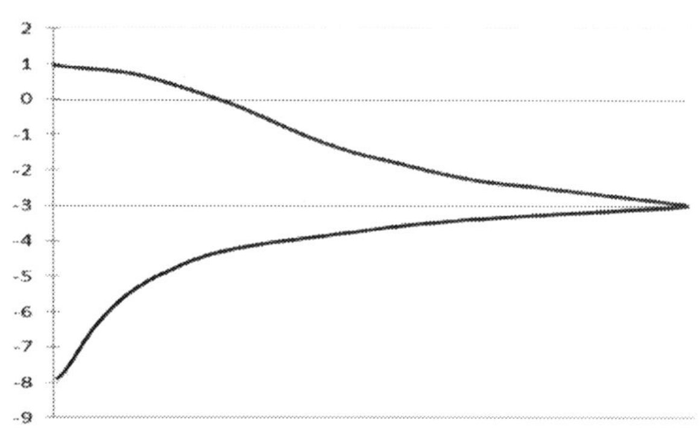

It is apparent that as victim's aggression rises, the payoffs for the sponsor reduce whilst the victim's Payoffs (while still remaining negative) increase. Extrapolation of the curves beyond the equilibrium point would logically follow the trend of increased payoff for Player A (terror victim) and associated increasingly negative payoff for Player B (terror sponsor) if it is accepted that A, being the stronger of the two adversaries, can inflict more damage than B even in case of the worst case scenario of a full-fledged war.[38] This 'extrapolation' could equally be studied by fixing the Nash Values with disparate values. This uses the logic that the equilibrium point is bad for both the players, but is substantially worse for the weaker adversary. The resulting curves are depicted below.

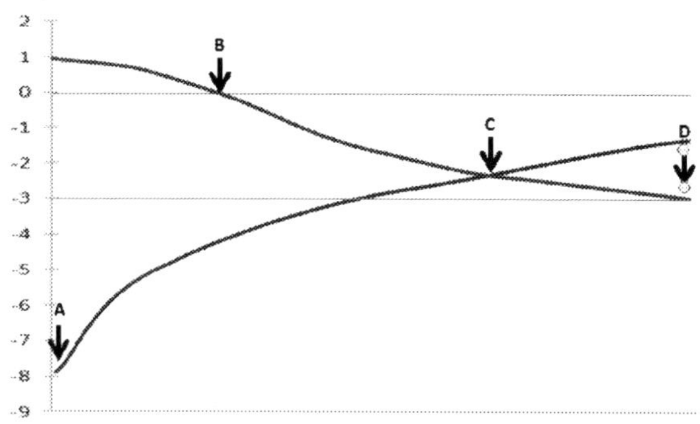

The following may be surmised:

(a) Points 'A' and 'D' on the curve represent the extreme cases where the game may be largely considered equivalent to its binary form.

(b) Point 'B' indicates where sponsor state's payoffs transition from the positive to the negative values. In other words while the weaker state is sponsoring terrorism as (from its viewpoint) it sees some benefits in the process, at this point on the curve, the losses begin to outweigh perceived benefits.

(c) The area between 'A' and 'B' represents reduced losses for the victim state (reduction in number of Incidents?) whilst still providing the adversary with a face saving option of not being forced to take a decision that may result in an all-out conflict that they would possibly like to avoid more than the stronger state would.[39] The limited aggression till this stage suggests that any escalation of violence in further interaction may be blamed on this decision were it to be forced. Victim's actions that hurt the sponsor, but not to the extent of forcing retaliation would come under this.

(d) Point 'C' represents a situation where the negative consequences for both the players are equal. Conventional deterrence may be assumed to have failed close to this stage. In other words beyond this point, the victim may be committed to aggression to a 'point of no return' and have nothing to lose with the situation escalating so long as the nuclear deterrence holds.

(e) The area between 'B' and 'C' represents the highest 'return on investment' for an aggressive victim viewpoint. This is therefore also the area where the strategic decision making should concentrate.

(f) Superimposing the troika of dissuasion, denial and threat on the curve between 'A' and 'C' suggests that deterrence by threat of punitive action with limited 'surgical' strikes to show resolve and make the threat credible in repeated games may be the only option closer to Point 'C'.

Having seen the theoretical underpinnings it may be useful to study the deductions against a real life incident such as the 26/11 Mumbai attacks. The obvious question that arises is that of the motive for the scale and the timing of the attack. Was it the intention to provoke India to attack? If so, what would be gained and was the Indian decision of military restraint the better option under the circumstances? What action must India take to prevent such strikes in the future and what action must be taken if such a strike were unfortunately to occur? All these questions are interlinked and have no easy answers.

The 26/11 attacks occurred at a time when political events in Pakistan were undergoing an upheaval. The internal situation was unstable. The year 2007 witnessed the Lal Masjid incident[40] and the formation of the Tehrik e Taliban[41] (TTP) headed by Baitullah Massod bringing together a reported 13 militant groups—some focused on challenging the Pakistani state. 2008 saw not only increasing targeted attacks on the Pakistani security forces (including on the Naval War College in Lahore and the Federal Investigation Agency Building) but also a 746 percent increase in attacks.[42] According to the conservative estimate 7997 people were killed and 9670 were injured.[43] On the political front this was juxtaposed with the General Elections of February 2008 which resulted in a coalition that had questionable ability to govern but appeared to desire an improvement in relations with India. President Zardari in his first address to the Parliament[44] (06 September 2008) sought improved relations with India and a solution to the Kashmir problem. He sought new initiatives to start cross LoC trade, more liberal visa regimes and full support to the Kashmiris for their 'fundamental rights' (as opposed to the previous stand of self-determination).

Even more seriously, the Pakistan president in an interview to the *Wall Street Journal*[45] on 04 October 2008 branded 'Jehadi' Groups operating in Kashmir as terrorists and compared them to 'wolf packs' who became active when India and Pakistan were talking peace. Just days before the attack, whilst addressing 'The Hindustan Times Leadership Summit' from his residence via satellite link, he spoke of shared bloodlines, a "bit of India in every Pakistani" and committed to no first use of nuclear weapons![46] With army being attacked from within and a civil leadership talking peace like never before, the cauldron boiled over. Nothing unites Pakistan like the hatred for India. Indian aggression at such a time, even in response to a terrorist strike by purportedly non state actors would possibly have suited the Pakistan army. Most commentators agree that the 26/11 attack was intended to provoke an Indian reaction.[47] Former ISI Director General Ahmed Shuja Pasha, speaking to the German media about the risk of war after the 26/11 Mumbai attack reportedly said: "At first, we thought there would be a military reaction. After the attacks, the Indians were deeply offended and furious, **but they are also clever**...(emphasis added)".[48] This may seem irrational given that the Pakistan army was already tied down fighting internal terrorism and would not therefore logically like to open another front. Facts about the existing situation however must be taken at face value. Unfortunately this again only leads to more unanswered questions. Even the most comprehensive studies into the incident have this trend of unanswered questions about the motives. Commenting on the work of Levy and Scott-Clark[49] in highlighting issues such as Pakistani army membership in LeT and Headley's relationships with Pakistani intelligence, Shamila Chaudhary lamented[50] that they "do not offer deeper insights into the nature of these relationships. For example, which parts of the Pakistani military deal with LeT? How high up the chain of command do the relationships go? Does LeT ever resist Pakistani directives? The reader becomes more aware of these problems but not much smarter as to their answers".

Insofar as the Indian response in concerned, it was largely restrained. In any case, if provoking India to attack was the aim of the operation, in hindsight the Indian restraint was arguably the better option. This restraint, which has been described as 'rage on a leash',[51] was limited to the military realm and other actions lower down the aggression scale were certainly initiated. The third necessary condition for stopping state sponsored terrorism, namely holding the sponsor state accountable was clearly exercised. The international community also played its part as seen from the number of high profile US visits to India and Pakistan and action taken by some other countries in the subsequent years. The US 'Enhanced Partnership with Pakistan Act' of 2009 (also known as the Kerry-Lugar-Bergman Act), while increasing the non-military development aid to Pakistan inserted clauses making this conditional on civil control over the military as well as action to curb terrorism from Pakistan soil. The terms of the act, agitated many in Pakistan including the Pakistani army. Following a meeting of its corps commanders, the army expressed "serious concern" over what it said were the "national security" implications of the aid package.[52] Clearly Pakistan was aggrieved and offended, but not offended enough to refuse the aid! This reality of the state of Pakistani economy is well known to both the US and Pakistan and must also be factored by India when deciding on further course of action. The US 'House of Representatives' Foreign Relations Committee' did not approve a further amendment which would cut all aid to Pakistan.[53] This was classical deterrence in that while future good behaviour was rewarded the 'power to hurt' was held in reserve and not actually used. The threat to cut aid was implied rather than stated. Any decision to actually cut all aid may have severely impacted the leverage that would otherwise be available in the future.

Finally, it needs to be examined what can be done to prevent a recurrence of a large scale terror strike akin to the 26/11 Mumbai attack. While planning deterrence for the future it must first be acknowledged that neither the sponsor state, nor the individual groups supported by it are in a position to up the ante beyond terrorism. Conventional war is not an option that can be economically sustained. In the event of a limited Indian strike, post another incidence, public opinion in the sponsor country may force retaliation. This is almost certain. It would however be their calculation that the confrontation would be limited and that the international community will not allow an all-out war. International opinion in such a scenario would therefore be critical. The sponsor state would be certain to 'play victim'. Bogey of nuclear confrontation as well as deliberate posturing of irrationality and ambiguity may also be used to attain the objective of controlling the damage while attempting to save face. However intransigent positions the two sides may adopt, it is also however certain that any aggression by the victim state would impact actions in the future by raising the costs for sponsorship of terror.

Decisions about what actions can be taken are obviously in the domain of policy making and cannot be discussed in unclassified academic discussions. It is for consideration, however, whether a statement by India outlining the action that would be taken in case of any further attacks could help shape the adversarial policy making.

Escalatory 'tit for tat' spiral can only be avoided if both sides have a shared understanding of the circumstances. This necessitates conveying Indian intentions with clarity and plausibility. This may not be possible when any action is being initiated due to crisis constraints. Therefore the options could be discussed during the intervening period prior to the next provocation. In fact if correctly managed, it may ensure that triadic deterrence holds and that there would be no subsequent provocation.

NOTES

1. Chari, P.R., Cheema, Pervaiz I., and Cohen, Stephen P., *Four Crises and a Peace Process* (Washington: Brookings Press, 2007).
2. Dasgupta, S., and Cohen, S.P., "Is India Ending Its Strategic Restraint Doctrine?" *Washington Quarterly* 34.2 (2011): 163-177.
3. Rajagopalan, Rajesh, *Nonalignment 2.0: A Realist Critique of An Establishmentarian Perspective*, IDSA comment accessed on 12 Jan 2014 at http://www.idsa.in/idsacomments/Nonalignment2ARealistCritiqueofanEstablishmentarianPerspective_rrajagopalan_010512.
4. Schelling, Thomas C., *Strategy of Conflict* (New York: Oxford University Press, 1963), p. 13.
5. Schelling, Thomas C., *The Diplomacy of Violence* (New Haven: Yale University Press, 1966), pp. 1-34.
6. Joint Publication 1-02, US Department of Defense Dictionary of Military and Associated Terms.
7. Goertz, Gary, and Starr, Harvey, *Necessary Conditions: Theory, Methodology, and Applications* (Lanham, Maryland: Rowan & Littlefield Publishers, 2003).
8. Brodie, Bernard, "The Anatomy of Deterrence," *World Politics,* 11, No. 2 (January, 1959): 173-191.
9. Kahn, Herman, *On Escalation. Metaphors and Scenarios*, Hudson Institute Series on National Security and International Order, No. 1 (New York: Praeger, 1965).

10. Fearon, James D., "Rationalist Explanations for War," *International Organization* 49, No. 3 (Summer, 1995): 379-414.
11. Davis, Paul K. and Jenkins, Brian M., *Deterrence and Influence in Counterterrorism* (Santa Monica, CA: Rand Corporation, 2002), p. 1.
12. Hanovich, Robert L. Jr, *Fuzzy Deterence* at http://www.dtic.mil/dtic/tr/fulltext/u2/a530056.pdf accessed on 13 January 2014.
13. Hunt, Carl and Chesser, Nancy (Ed), *Deterrence 2.0: Deterring Violent Non-State Actors in Cyberspace*, Workshop held 9-10 January 2008 in Arlington, US Strategic Command Global Innovation and Strategy Centre.
14. Knopf, Jeffrey W., *The Fourth Wave in Deterrence Research* at http://www.contemporarysecuritypolicy.org/assets/CSP-31-1%20Knopf.pdf accessed on 13 January 2014.
15. Bunn, M.E. *Can Deterrence Be Tailored?* (Washington, D.C.: Institute for National Strategic Studies, National Defence University, 2007).
16. Atzili, Boaz and Pearlman, Wendy, "Triadic Deterrence: Coercing Strength, Beaten by Weakness," *Security Studies*, 21:2 (2012): 301-335.
17. Gareau, Frederick H., *State Terrorism and the United States: From Counterinsurgency to the War on Terrorism* (Atlanta, GA: Clarity Press, 2004), pp. 14-15.
18. Hoffman, Bruce, *Inside Terrorism* (Columbia University Press, 2006), pp 229–256.
19. Steed, Brian, *Piercing the Fog of War: Recognizing Change on the Battlefield: Lessons from Military History, 216 BC Through Today* (Minneapolis: MBI Pub. Co. LLC and Zenith Press, 2009), pp 251.
20. Payne, John D., *Fighting for Control: State-sponsored Terrorism as Foreign Policy in Cuba and Libya, 1959-2010, 2011* (Massachusetts Institute of Technology, 2011).
21. Ibid.
22. Polk, William R., *Violent Politics: A History of Insurgency, Terrorism, & Guerrilla War, from the American Revolution to Iraq* (New York: Harper, 2007), p. 18.
23. Gladwell, Malcolm, "How David Beats Goliath: When Underdogs Break the Rules," *The New Yorker 85*, No. 13 (May 11, 2009): 40.
24. Three models proposed by Zagare and Kilgour (2000) for perfect deterrence include unilateral deterrence, generalized mutual deterrence and asymmetric escalation.
25. Quackenbush, Stephen L., "General Deterrence and International Conflict: Testing Perfect Deterrence Theory", *International Interactions*, 36, 2010: 60-85.
26. Ibid. The details of the conventions used and the rationale for the conclusions are not included. The paper cited may be referred for further details. The limited point sought to be made is that the number of possibilities can be substantially reduced if the basic assumption of the state sponsoring terrorism being constrained in openly initiating conflict is taken to be true.
27. Myerson, Roger B., *Force and Restraint in Strategic Deterrence: A Game-Theorist's Perspective* (Carlisle Barracks, PA: Strategic Studies Institute, U.S. Army War College, 2007).
28. Ibid.
29. Ibid.
30. 'Strategy' in a game is the complete plan that specifies an action for the player in

every possible situation that the player could encounter in the game.

31. Roberts, Brad, *Deterrence and WMD Terrorism: Calibrating its Potential Contributions to Risk Reduction* (Ft. Belvoir: Defense Technical Information Center, 2007), p. 275.
32. Jackson, Brian A., *Evaluating Novel Threats to the Homeland: Unmanned Aerial Vehicles and Cruise Missiles* (Santa Monica, CA: RAND National Defense Research Institute, 2008), p. 95.
33. Payne, Keith B., *Deterrence in the Second Nuclear Age* (Lexington: University Press of Kentucky, 1996), p. 129-32.
34. See Pape, *Coercion and Military Strategy*; Byman and Waxman, *Dynamics of Coercion*; Brodie, *Strategy in the Missile Age*.
35. Bertrand Russell, *Vagueness*, 1923, accessed at http://vserver1.cscs.lsa.umich.edu/~crshalizi/Russell/vagueness/ on 14 February 2014.
36. Rules are usually expressed in the form: *IF* variable *IS* property *THEN* action.
37. Boolean operators do exist in fuzzy logic but have different connotations. The AND, OR, and NOT operators, for example, are usually defined as the minimum, maximum, and complement; being termed Zadeh operators when defined this way. Other operators, more linguistic in nature, called hedges can also be applied. For the fuzzy variables x and y:
    NOT x = (1 - truth(x))
    x AND y = minimum(truth(x), truth(y))
    x OR y = maximum(truth(x), truth(y))
38. For the purpose of the discussion only an all out conventional war is assumed to be the worst case scenario. Nuclear deterrence between the two players may be seen to be working separately and is assumed to hold.
39. This would be the logical assumption based on the three conditions for state sponsored terror discussed earlier in conjunction with Pakistan being viewed as the weaker state.
40. The Siege of Lal Masjid code-named Operation Sunrise was a confrontation in July 2007 between Islamic fundamentalist militants and the Government of Pakistan led by General Pervez Musharraf and Shaukat Aziz Administration, then Prime minister of Pakistan. The focal points of the operation were the Lal Masjid ("Red Mosque") and the Jamia Hafsa madrasah complex in Islamabad, Pakistan. See http://en.wikipedia.org/wiki/ Siege_of_Lal_Masjid (accessed 16 February 2014) for detailed citations and further information.
41. Among the Tehrik-i-Taliban Pakistan's stated objectives are resistance against the Pakistani state, enforcement of their interpretation of sharia and a plan to unite against NATO-led forces in Afghanistan. See http://en.wikipedia.org/wiki/Tehrik-i-Taliban_Pakistan (accessed 16 February 2014) for detailed citations and further information.
42. As compared to the figures for 2005.
43. *Pakistan Security Report 2008*, Pakistan Institute of Peace Studies.
44. President Asif Ali Zardari's First Speech to the Parliament Promises, Commitments and Actions, accessed at www.cpdi-pakistan.org on 16 February 2014.
45. Audio clips available at http://online.wsj.com/news/articles. The statements led to

protests and issue of a disclaimer by information minister Sherry Rehman reiterating Pakistan's continued diplomatic support for the Kashmiris.

46. Sharma, Vinod and Zia Haq, "There is a bit of India in every Pakistani: Zardari," *Hindustan Times*, New Delhi, 22 November 2008.
47. See, for example, "26/11 attack intended to provoke India-Pakistan war: US expert," News article in the *Economic Times*, accessed at http://articles.economictimes.indiatimes.com/2013-11-19/news/44242435_1_mumbai-attack-david-headley-terror-attack on 12 Mar 2014.
48. Siddiqa, Ayesha, "Indo-Pak Conflict and the National Egos," accessed at http://www.tehelka.com/street-aggression-helps-little-in-solving-indo-pak-problems/ on 12 Mar 2014.
49. Scott-Clark, Cathy and Levy, Adrian, *The Siege: Three Days of Terror Inside the Taj* (Viking, 2013).
50. Chaudhary, Shamila N., "Why India and Pakistan Hate Each Other, The South Asia Channel," accessed at http://southasia.foreignpolicy.com/posts/2014/02/04/why_india_and_ pakistan_hate_each_other on 12 Mar 2014.
51. Reyar, Akrita, "India-Pakistan on a road to nowhere," accessed at http://zeenews.india.com/MumbaiAttacks/ story.aspx?aid=579906 on 12 Mar 2014.
52. Waraich, Omar, "How a U.S. Aid Package to Pakistan Could Threaten Zardari," *Time Magazine*, 2009 accessed at http://content.time.com/time/world/article/0,8599,1929306,00.html on 12 Mar 2014.
53. For a detailed discussion on US aid to Pakistan, legal provisions and waivers etc see Susan B. Epstein and K. Alan Kronstadt, Pakistan: U.S. Foreign Assistance, CRS Report for Congress, 2013.

CHAPTER 10

Ransom Demands by Pirates: Study of the Problem as a Bargaining Game[1]

Piracy, in its essence, is an economic transaction involving purely monetary considerations. This immediately sets it apart from terrorism, which may be rooted in ideology or religious beliefs. Even if one were to accept that the roots of piracy off Somalia lie in the sense of injustice perceived by the local fishermen (due to poaching by extra regional trawlers or dumping of toxic wastes which directly impacted their livelihood etc), it must be acknowledged that the pirates are essentially looking for money that can be obtained as ransom for the hijacked ship and its crew.[2]

Some of the existing economic models can therefore be directly applied to the transaction of ransom demands by the pirates. The ransom negotiation process is thus similar to any bargaining game, where each side is trying to negotiate terms favourable to itself. This process of negotiation is very similar to the negotiations that may be undertaken by commercial firms, trade unions etc., even though there would be some peculiarities to each situation.

A negotiation has been defined as a "joint decision-making process of two or more parties working together to reach a mutually acceptable agreement over one or more issues".[3]

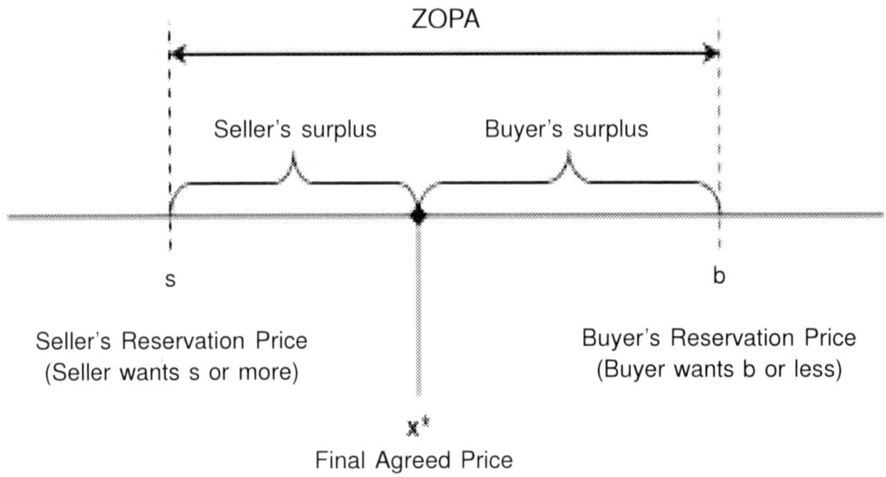

Figure 10.1: Zone of Potential Agreement

Any negotiation process normally involves at least two parties. Each of these stakeholders would have a 'limit' on what is acceptable. This limit forms the basis of all negotiations. Even a simple commercial transaction such as buying a house, for example, may be studied from this perspective. The seller would have a figure which is considered the lowest acceptable. The buyer would similarly have placed a value corresponding to the highest that can be paid. The overlap between the two is the area in which the final agreement may be reached. There thus exists a Zone of Potential Agreement (ZOPA), as depicted in the figure 10.1.[4] The negotiation process is about each party trying to shift the final agreed price closer towards the opponent's limit and thereby maximizing the 'surplus'. The agreement, it must be noted, can only be reached within this ZOPA. If the lowest acceptable is higher than the highest that can be paid, there would be no overlap and there would be no transaction. The problem, when negotiating, is often of incomplete information. Each side does not know what the other side has determined as their limit or reserve price and can at best make an 'informed guess' about this aspect. Prevalent market price, availability of other options (say competitive sellers) etc can influence the estimate about the other player's limit as well as be a factor in setting the limit itself.

Another problem that may be encountered is that the limits may vary over time, thus leading to a variable ZOPA. This may result from a number of factors. It could be due to the negotiation process itself or due to other factors such as depreciation of the property over the period of prolonged negotiations.

The commercial transaction of purchasing property, with minor modifications, may also be suitable for studying the economic transaction that the pirates and the ship owners indulge in. The pirates would have a lower limit below which they would not be willing to transact and the ship owners would similarly have an upper limit on what they are willing to pay (or capable of paying). Unlike a transaction of buying a house, however, the ransom situation has some peculiarities. It may be described as a "mutually hurting stalemate"[5] if both parties believe that the status quo is not satisfactory and that an agreement will benefit everybody.

Some commentators have described hostage negotiation in the pirate's scenario as a "bilateral monopoly price negotiation that is structurally just a special case of chicken".[6] The rationale being that, unlike a "barrel of oil or a freight car full of soybeans which can trade on an extremely liquid market with innumerable buyers and sellers, a hostage has exactly one seller (the kidnappers) and exactly one buyer (the employer and/or family of the hostage). When there is only one buyer, the opportunity cost for ransoming the hostage is zero. Likewise, the employer and/or family has no realistic alternative means to recover the hostage. In order for everybody to walk away happy, we need a cooperate-cooperate outcome: the kidnapper has to give up the hostage and the employer/family has to give up a ransom". If this view is accepted and we model the situation on a bilateral monopoly negotiation only two things should matter.[7]

- The first is the willingness to accept failure. The more willing you appear to terminate the negotiation process, the more bargaining power you have.
- The second is the capacity to pay. The pirate's demand, even in a monopoly situation, is pointless if it is beyond the capacity of the other party to pay.

As per this view, the 'going rate' does not matter. This is based on the theory that 'market price' is irrelevant to a bilateral monopoly as there is no market but only a series of discrete one-off transactions. In the absence of substitutability, 'comparable' transactions are irrelevant. In normal markets the going rate matters, but only because it provides the opportunities for substitutes and this creates the 'law of one price'. It is "completely irrelevant that in the past kidnappers accepted ransoms of $x/2 since I don't have the relatively good fortune of dealing with a kidnapper who demands $x/2 but am stuck with one who demands $x".

The view articulated does however accept that there is an issue of information asymmetry in that the hostage's party has much better information on its assets than do the pirates and so the pirates may be sceptical of the hostage's party pleading poverty. It can therefore be argued that lack of complete information makes the pirate bargaining dissimilar to the simple case of bilateral monopoly. This lack of complete information exists not only for the pirates but equally for the ship owners.

In studying the piracy related ransom demands as a bargaining game, the first stark difference as compared to a commercial transaction considered above, is the variable ZOPA. Since the value of the commodity (that is the ship) varies over time and the pirates themselves incur continued costs during the prolonged negotiations, the ZOPA is obviously not constant from the perspective of either party. At first look, the ZOPA narrows over time as the costs of the pirates rise (to guard and take care of the hostages) and the value of the ship for the owner depreciates. Further, the hidden costs, over and above the ransom, have to be factored in. As per one estimate,[8] "The professional negotiators, acting on behalf of the ship owners, get about $100,000 for their services and the lawyers receive a fee of about $300,000 for ensuring that the shipping companies are not putting themselves in any dubious positions". The cost of transferring the ransom (This may involve up to two aircraft) and interests on any loans also have to be factored in.

A detailed study[9] on the negotiation process by EUSECON (European Security Economics), a research project supported by the European Commission's Seventh Framework Programme is another study that uses Game Theory based examination of the problem outlined above. The first thing that determines the final ransom paid, is not the initial demand or even what the ship owner is willing to pay, but rather what the pirates are willing to accept. What they 'accept' is certainly linked to what they 'expect'. Let us therefore first assume that there is a 'going rate' that is used as a benchmark by the pirates. This bench mark is termed as the Reference Level (RLp), with the letter p signifying the pirates perspective. A number of factors can determine what the Reference Level in any situation would be. The tonnage and age of the ship, cargo carried and the flag state are the most obvious. In addition the number and nationality of the crew may be an important factor. Crews from rich nations are likely to fetch a higher ransom.[10] Once a reference level is assumed, it is apparent that a ransom offer R, such that R > RLp would be considered satisfactory by the pirates and the hostage situation would be satisfactorily terminated. A ransom R < RLp would, on the other hand, be considered unsatisfactory and will result in further negotiations.

In addition to the obvious factors already discussed, the additional primary factor, as far as the pirates are concerned, would be the cost incurred by them in capturing the ship. We designate this as the Cost of Operations or C_0. The readily measurable monetary costs (pay per pirate, cost of weapons, operating costs for the mother ship, cost of skiffs and OBMs, communication equipment, provisions) as well as other factors such as failed attempts, pirates arrested or killed in naval operations etc would all contribute to the overall cost. Since the latter factors would reduce the success rates and every mission may not culminate in a successful capture of a ship, the C_0 may actually be the cumulative cost of many missions. The anti-piracy measures including those by naval forces, private security companies and the Best Management Practices followed by ships, also therefore contribute to the enhanced cost of operations from the viewpoint of the pirates.

The ransom that the ship owner is willing to offer would similarly depend on a number of factors. The primary factor determining what is offered would be the Value of the ship (*v*) which is assumed to include the value of the cargo on board. This value is again not fixed and is difficult to quantify. We assume that the value '*v*' is the initial value which will depreciate over a period of time as the cargo may perish, the ship may suffer material deterioration due to lack of maintenance etc. For each time period 't', the value is therefore assumed to depreciate by a factor θ.

As the negotiations are progressed, the costs for both the players go up over time. The ship owner loses out not only charter income from the ship, but also spends money on conduct of the negotiations. Interest on any loans that may have been taken and remuneration to the captive crews' families are also factors. The pirates similarly have to pay not only for sustenance of the crew but also for guarding them and the ship as well. Negotiations themselves cost money, even for them. The cost of prolonging the negotiations will not be the same from the perspective of the ship owners and the pirates since the factors involved in each case are different. We designate C_s and C_p be the per-period cost of prolonging negotiations for the ship owner and pirates respectively.

Further, in addition to these costs, both parties would be keen on solving the impasse quickly due to the risks associated with prolonged negotiations, some of which may become critical. As for example the pirates' ability to retain control of the vessel in face of threats by rival gangs or the ship owner's time-to-insolvency considerations would be included in this. The expected payoffs must be discounted by a factor which is indicative of these risks. Future payoffs are accordingly discounted by a factor δ_p by the pirates, and by a factor δ_s by the ship-owner, both of which may vary between (0,1).

Once the negotiations start, the players would bargain as per a sequence of alternating offers. If the pirates make the initial offer, the ship owner would make a counter (lower) offer after a time-unit. The third offer would be made by the pirates, fourth by the owner and so on. The actual negotiation process may deviate from this strict alteration, however the fact that the lower or higher of two simultaneous offers may be taken as indicative ensures that negotiation will, in essence, be a series of alternating offers. Each subsequent offer made by the pirates can be expected to progressively come down as the previous higher offers are rejected. The attempt during negotiations would be to make the owners increase their offers so that an agreement can be reached. The owners would similarly be trying to lower the demand. In each period $t = 1,..., \varphi$, a Ransom offer, R_t, would thus made, and would either be accepted or rejected. This sequence of offers would end with an offer being accepted. The instantaneous payoff from accepting offer R_t at time t may be quantified as:

$$\Pi_{pt} = R_t - RL_P - C_o - tC_p \qquad \text{for the pirates,}$$

and $\Pi_{st} = \theta^{t-1} v - R_t - tC_s$ for the owners

It is also possible that no agreement can be reached. If the negotiation continues till the final period ($t = \varphi$), and the final offer is rejected the pirates still have the ship, which has some intrinsic value (it could for example be used as a mother-ship in future expeditions). The payoff in such a case will therefore be the Present Value (V_P), with $V_P > 0$. The ship owners will similarly be able to recover some amount through an insurance claim (V_I). The terminal payoffs are therefore,

$$\Pi_{p\varphi} = V_P - RL_P - C_o - \varphi C_p \quad \text{for the pirates,}$$

and $\Pi_{s\varphi} = V_I - \varphi C_s$ for the owners

The final outcome of the bargaining when an agreement is reached is given by a ransom r, and a duration d. The prediction of these using Game Theory is given by the sub-game perfect equilibrium of the game. This may be found by backwards induction from the last period. If the ship owner moves last[11] ($t = \varphi$) the pirates must be offered at least V_P, so $r\varphi = V_P$. For any other value of $t < \varphi$ the equilibrium requires that the offer made must equal an amount such that the other player is indifferent to accepting the offer immediately and accepting a counter offer after a one-period delay. Hence the sequence of offers of the pirates satisfy

$$\theta^{t-1} v - R_t - tC_s = \delta_s (\theta^t v - R_{t+1} - [t+1]C_s) \qquad t = 1,3,\ldots, \varphi-1$$

The offers of ship owners similarly satisfy

$$R_t - RL_P - C_o - tC_p = \delta_s (R_{t-1} - RL_P - C_o - [t-1]C_p) \qquad t = 2,4,\ldots, \varphi-2$$

Since both the players suffer diminishing payoffs as the time progresses, the mathematical solution of such a game under conditions of *complete information* would be predicted to end after the first offer (d=1) and has no bearing on negotiation length. The fact that this is practically not observed to be true is reflective of the fact that both the parties are negotiating with incomplete information. It is therefore important to analyse how the information is incomplete and the effect on the negotiation process.

The pirates may be classified as 'amateurs' or 'sophisticated'. The former are opportunistic and do not have the complex support infrastructure (including shore based support infrastructure) that the gangs of sophisticated organized crime syndicates do. Obviously the cost involved in the operations, the expected payoffs as well as the time that they can prolong the negotiations are higher in case of sophisticated pirates. The owners may similarly be classified as 'rich' or 'poor' ranging from big shipping companies to owners of small fishing trawlers. Except in the case of well known shipping companies, it would benefit the owners to pretend to be poor and plead lack of funds. The pirates would similarly pretend to be sophisticated. The time for which amateur pirates can engage in negotiations (φ_a) is small as compared to f_s for sophisticated pirates. If δ_s is sufficiently high, the ship-owner would extend the negotiations so as to reveal the pirate's type. Amatuer pirates in such cases would give in and accept a lower ransom. If the negotiation extends further, the information asymmetry becomes one sided. The owners know that the pirates have the capability to extend the negotiations while the pirates still do not know the owner type with certainty. In such a case it would be in the owner's interest to end the negotiations quickly, thus resulting in a higher ransom being paid. The owners incapable of doing so must wait till the time the pirates are convinced of their inability for the ransoms to dip again—thus leading to a hump shaped curve when plotting the ransom against time.

Whilst the study cited ibid captures the dynamics of a pirate negotiation well from the point of view of the ship, the process of trying to game the situation from the point of view of the hostages is slightly more complicated. In extending the negotiations, which may be the logical course of action for certain pirate/owner category, the agony of the hostages or the family members is overlooked. In addition, there is always the danger that the hostages may die in captivity or indeed be killed by the pirates in case their demands are not met. Taking a broader view, the ship and the hostages may therefore be seen as separate as far as the bargaining is concerned. As far as the ship is concerned, the ship owners lose money every day in case of prolonged negotiations causing unavailability of their vessels for charter. The deterioration in the condition of the ship over time itself also affects this depreciation over time. From the point of view of the hostages, the owners would face increasing pressure (political as well as from the hostages' families) over time to reach an agreement. If a hostage is killed or injured, the vessel owner or the shipping company may have to, in some countries, stand trial or be sued. Similarly, the pirates would be willing to accept a lower ransom if they are convinced that a higher payment is not forthcoming.

This has been covered below using a simple model developed by Reinhard Selten in 1976. While the original model dealt with a kidnapping scenario with the danger of the Kidnapper being identified by the hostage leading to the eventual arrest being modelled against the threat of killing the hostage, it can be suitably modified to the pirate ransom situation. The game has been kept simple with a minimum number of variables in order to try and understand the responses of the pirate if the expected ransom is not forthcoming despite prolonged negotiations.[12] The game would include the Pirates (P) and the Ship-owners (S). It is assumed that the negotiation process has reached a stalemate where S is willing to pay a Ransom amount C, while P demands a minimum sum R_p.

For the stalemate to continue, C < D.

The pirates would have incurred some costs for the operation and this would determine the ransom D demanded by them. They would likely threaten to kill the hostages if the demand is not met. While they have nothing to gain from killing the hostages, they may carry out the threat to make their subsequent threats credible[13]. Let us quantify the probability of their carrying out the threat as á.

If C – D, the probability is the minimum $\alpha = 0$

If C = 0, the probability is the maximum $\alpha = 1$

For the sake of simplification, we assume that α can thus be described as a linear function (C/D)

$$\alpha = a\,[1-(C / D)] \text{ for } 0 \leq C \leq D \qquad \text{(i)}$$

where a is a constant such that $0 < a < 1$.

Other than non-rational execution of threats due to factors such as frustration,[14] the decision to carry out the threat can be modelled on a binary decision variable e such that

e = 0, when the offer C is accepted and the hostages released

or e = 1, when the threat is executed.

It is assumed that the killing would lead to some negative publicity for the pirates and may also result in international action against them being intensified. We quantify the probability that action against the pirates would be intensified as q such that

$$0 < q < 1$$

Even though it would be difficult to quantify the payoffs in such a scenario, we assume that an action would not be undertaken unless there was something to be gained from it. We therefore assign positive constants w, x, y, z to the payoffs for different situations without attempting to quantify them other that the primary assumption that they are all positive. The payoffs for the adversely affected player would be denoted with a negative sign. For example in the eventuality that the hostages are executed the payoff for the owners is (–w) regardless of what happens to the pirates thereafter. If the hostages remain alive though the ransom offered is not accepted the payoff for the owners is still negative (–v) though ($v < w$) as there may still be hope for subsequent release or rescue. The pirates will also suffer a negative payoff (–x) in such a scenario which they must weigh against the possible benefits of keeping the hostages alive (possibly for use as a human shield or crew for mother ships as has also been observed on some occasions). If the hostages are executed and yet the pirates are not subject to any additional action, the pirates gain, thus $z \geq x$. The expected payoffs are tabulated below:

	Payoffs	
Outcome	*Pirates (P)*	*Ship-owners (S)*
Release of hostages on accepting C	C	–C
Status quo	–x	–v
Execution of hostages, international action against pirates remains same	–y	–w
Execution of hostages, international action against pirates intensified	–z	–w

The same can be modelled as an extensive game with perfect information, wherein at every point the players know the complete previous history leading up to the event. The 'optimal' choices may be calculated if we can uniquely determine an equilibrium point in any situation.

Optimal Choice of 'e'. For the sub-game which begins with the pirates deciding to carry out the threat, let the payoffs be K_1 and K_2 respectively for 'not killing' or 'killing' the hostages.

$$K_1 = (1–q)\ C – qx$$

$$K_2 = –\ (1–q)\ y – qz$$

Since $C = 0$, $y > 0$, $z = x$ and $(0 < q < 1)$, we have $K_1 > K_2$ in all conditions. Therefore the option of 'not killing' is always the rational choice and e=0 is the optimal solution.[15]

Optimal Choice of 'C'. For the sub game which begins with the ship-owners choice of C, even if player S realizes that P will choose $e = 0$,[16] the probability of the pirates carrying out the threat is still α. The expected value of his utility would be:

$$U = –\ (1 – \alpha)\ (1 – q)\ C – \alpha\ w$$

Substituting the value of α from equation (i) and rearranging terms in quadratic function[17] of C

$$U = \{–a\ (1–q)/D\}\ (C)^2 + \{[aw/D] – (1–a)\ (1–q)\}\ (C) – aw$$

In order to determine the optimum value of C we compute $\delta U/\delta C$

$$\delta U/\delta C = \{–2a\ (1–q)\ (C/D)\} + \{(aw/D)\quad [(1\ \ a)\ (1–q)]\}$$

This assumes the maximum value

$$C = w/[2(1–q)] – D\ [(1–a)/2a] \quad \text{if C is in the interval } 0 \leq C \leq D.$$

We next take the two critical values of D, with values D_1 and D_2 such that,

$$D_1 = aw/(1+a)(1–q) \text{ and } D_2 = aw/(1–a)(1–q)$$

For $D < D_1$ the derivative $\delta U/\delta C$ is positive in the interval $(0 \leq C \leq D)$. For $D > D_2$ the derivative $\delta U/\delta C$ is similarly negative in the given interval. In this case the optimal value of C' becomes

$= D$	for $0 < D \leq D_1$
$= w/[2(1–q)] – D\ [(1–a)/2a]$	for $D_1 \leq D \leq D_2$
$– 0$	for $D \geq D_2$

In simple terms, with an increase in the ransom demanded D, the optimal offer C first increases up to $D = D_1$ and then decreases until it becomes 0 at $D = D_2$. Up to the lower critical value D_1, S eliminates the danger to hostages by yielding to the demands of P. In the interval between D_1 and D_2, the probability of the pirates carrying out the threat α is diminished with additional money C, with the change being lesser for higher values of D. After the higher critical value D_2, influence on á is so small that it appears useless to offer anything at all and the negotiations are likely to be terminated.

Optimal Choice of 'D'. We next look at the sub game which begins with the pirates choice of D. If the pirates know that the player S will select their offer optimally and that later they themselves will choose e=0, the payoff expectation may be calculated under these conditions. We have seen that the payoff as a function of D increases up to D_1 and then decreases up to D_2. It therefore follows that the optimal value of D would be at D_1. Players P's optimal demand would be when the optimal offer coincides with the demand or when

$$D' = aw/(1+a)(1-q)$$ [18]

Observations and Policy Recommendations

It may therefore be seen that both the ship and the hostages are amenable to being used for bargaining though the modalities governing the negotiation process are different. In case of prolonged negotiations, as the cost of the pirates mount, there are often disagreement between the pirates on the course of action. Jon Lee, an analyst for Compass Risk Management company, which has been involved in negotiations to free ships and crew, "Pirates treat the crew and the ship as commodities and they have a very businesslike attitude about ransoms and hostages".[19] In one case the crew and the ship were split as the 'financier' refused to release a part of the crew even when the ship had been ransomed. "In this case, the commodity has been split between different gangs. Each group will be looking at how much the other got. The first group does not want any other group to get more than they did and the amount they obtained will govern and influence how much the rest ask for".[20]

It has often been stated that piracy off Somalia is on the decline. This is apparent by the number of attacks over the last few years. The decline is attributed to a number of factors including patrol by naval forces, adoption of Best Management Practices, presence of private security companies etc. From the purely economic viewpoint this apparent decline is arguable as only statistical figures indicating a decline in the number of successful or attempted attacks over a period of time may be insufficient to conclude effectiveness of these measures unless a consequent decline in the 'profits' is also observed. The measures indicated certainly act as deterrents in that they increase the Cost of Operations for the pirates. If they do not succeed sufficiently in checking even the isolated success rates, the failed attempts just go into increasing the Reference Level for the subsequent ransoms thereby driving them up. One large successful attempt may economically be a better option for the pirates, than a number of smaller ones. There is no doubt that the number of incidents has declined in 2012. The reasons for the same however may be other than those claimed, and the economic model continues to be in place. About the reasons for the decline IHS Jane's quoted[21] EUNAVFOR officials as stating, "We're not really sure why that is—we would like to think that we're doing something right and that the navies have had some effect on that. However, there are other factors at play: have they just got enough money? Is there something else going on in Somalia? We're really not sure, although we hope we've at least had some effect on it." The realization that the effect may be due to simple economic considerations is also there, "So this [statistic] needs to be taken with a grain of salt. We think that possibly one of the factors in the different pirating activity was the fact they were so busy focusing on getting money for the ships they had ... they basically ran out of capacity to hold onto ships, so they focused on ransoming them, getting the money and then once they got rid of their current inventory they could go about finding new inventory."

The question that must be answered is whether these 'deterrents', have ceased to have a deterrent effect. In addition to driving the ransoms up they also place the governments/ship-owners in a scenario where the costs are incurred even when there is a decline in the number of attempts. Since higher past ransoms are positively associated with subsequent ransom amounts, they also impart a negative externality on future victims. This in effect means a catch 22 situation where while these actions were ineffective to begin with, their termination would make things even worse. In such a scenario, the incidents as well as the ransoms would see an upward trend.

It is also important to note that ransom durations have lengthened as pirates have developed the land-side infrastructure needed to protect and supply ships for long periods of time. If the finding that sophisticated pirates extract higher ransoms than amateur is true, the increase in the ground based infrastructure to sustain long negotiations means that the negotiations will progressively be longer and the ransom higher. The decrease in the number of incidents notwithstanding, the effect of piracy continues to grow from the pirates' perspective.

The pirates have also shown a remarkable ability to adapt to the changes brought about by new security measures. While the presence of armed security guards has been quoted as a reason for the decline in piracy, the same also needs to be examined further. The pirates have shown themselves to be risk averse as is also shown by the decline in piracy incidents in the monsoon months. If there are enough ships available without the armed guards, it makes little sense to escalate the level of violence by engaging in a direct fight with the guards. This explains why not a single ship with guards has ever been attacked. The presence of guards on all ships may however result in the pirates having no option but to escalate the level of violence. This enhanced 'cost' may also result in a consequent increase in the reference level for ransom. The situation is akin to the pirates being forced to expand their area of operations based on the establishment of the international transit corridor. The change of tactics, necessitated by the security forces action, resulted in the use of mother ships and spreading the menace which had been a limited phenomenon till then.

It has been advocated that control of piracy can only be achieved by action on land. The conditions in Somalia where there is complete anarchy and the lack of other alternatives is cited as the reason for the former fishermen taking up piracy. Even if this view is accepted, it is doubtful if creation of jobs/infrastructure ashore will result in cessation of piracy. It must be acknowledged that anarchy does not favour piracy as the capacity of the pirates to prolong negotiations is directly affected. Also areas in Somalia which are relatively well governed have higher rates of support to piracy. While the country may not have a formal governance structure, the informal economy has been doing reasonably well, at least in the piracy affected areas. The change in economic conditions must therefore be seen as being independent of the piracy phenomenon.

The likelihood of the pirates rationally killing any of the hostages is almost nil. The only reported case of negotiations breaking down so as to involve killing as a threat was when Somali pirates who have been holding a hijacked ship for nearly two years reportedly killed a Syrian crew member[22] and wounded another to protest delayed ransom payment. Hassan Abdi, a pirate commander in Haradhere town, a key pirate centre, said that the killing was a message to the owners of the ship MV Orna that was hijacked off Seychelles in 2010 adding "More killings will follow if they continue to lie to us—we have lost patience with them. Two years is enough". This was subsequently found to be untrue. As per Abdul Kadar, technical adviser to Sirago Ship Management, which handles operations for Kasaab Intershipping, a company registered in the Marshall Islands, "It was a rumour but no one has been killed".[23] He added that the freed crew and the pirates had confirmed that all 19 crew members were still alive, though six crew member had not been released due to differences between the pirates. Other killings may have been as reprisal for attacks on the pirates by a particular country (for example the reported killing of four American hostages[24]). Since two dead pirates were also found on the yacht, it is more likely that the killing was accidental or during a failed attempt to overpower the pirates by the hostages. The attempt to spread fear using mind games is however apparent. Similarly, some Indian crew members' hostage ordeal is being prolonged reportedly in retaliation for the arrests of more than 100 Somali pirates by the Indian Navy.[25] This may also be a ploy to cash in on the information asymmetry as other Indians on the same ship were released.

Based on the above, the policy recommendations that emerge are summarized in the succeeding paragraphs.

- The role of the Government should be largely passive in the negotiation process as active involvement changes the information advantage in favour of the pirates—the owners can no longer claim to be poor. If the governments cave in to the political pressure the ransom is likely to be comparatively higher. For instance, "under political pressure to recover hostages, the Spanish government paid 1.2 million USD in 2008 for release of the Playa del Bakio, more than twice the previous record amount for a fishing vessel".[26]

- Since the safety of the crew can be assumed, prolonging the negotiations to help identify the amateur gangs from the sophisticated may be an option. Government involvement to help facilitate this aspect may be considered.
- 'Catch and Release' policy being followed by certain countries does not appreciably raise the cost of operations whilst still increasing the expected ransom. This may therefore be more harmful than no action.
- The need for cooperation amongst countries, especially in offensive actions, is apparent. This would not only prevent crew members from a particular country being specifically targeted but also substantially increase the cost of operations for the pirates.
- Escalation or threat of escalation in the action against pirates is the only credible factor to lower the menace. The action by Indian Navy in controlling the spread of piracy close to Indian shores is one example of proactive action being the optimum solution. Action to target, shore based logistics support depots immediately after an act of piracy may provide one option for raising the 'cost' for the pirates. Most countries are reluctant to get involved 'on ground' in Somalia, however targeting infrastructure close to coast which may not involve human life may be an option.

Capt Mukundan of IMB has pointed out that most piracy related costs are recurring in nature.[27] "What we [at IMB] believe is that a lot of action has been taken already at sea against the pirates. The report by the One Earth Future Foundation in 2011 estimated that last year the economic cost of piracy was between US$ 6.6 billion and US$ 6.9 billion. What is significant is that they estimated that 80 per cent of this cost is a recurring cost which we can expect to incur next year, the year after and so on. So we have reached the stage where we need to have a game changer, otherwise we can expect to carry on at the similar levels of piracy for years to come," he said. The need to seek changes to the current model to ensure that his grim prediction does not come true will need international cooperation and resolve to stop the menace. Targeting the economic model which supports piracy may be one, and possibly the only, long term action.

NOTES

1. Some of the aspects discussed here were initially published by the author in the article "Ransom Demands by Pirates: Study of the Problem as a Bargaining Game," *Maritime Affairs: Journal of the National Maritime Foundation of India*, 9:1, 110-128, Routledge, 2013.
2. The closest Somali translation for pirates is 'ocean robbers ('burcad badeed'). Somali pirates prefer to call themselves 'badaadinta badah', 'saviours of the sea': the movement started when Somali fishermen decided to react against illegal fishing and toxic dumping in Somali waters. However, the same is irrelevant as over time the problem has moved on from fishermen 'gathering taxes' from foreign vessels to a highly organised hostage taking business. It may be seen that fishing vessels though easier targets are no longer the preferred choice of targets. See also Stig Jarle Hansen, Debunking the Piracy Myth: How Illegal Fishing Really Interacts with Piracy in East Africa, *RUSI Journal*, December 2012, Vol. 156, No. 6.
3. Cohen, Steven, *Negotiating Skills for Managers* (New York: McGraw-Hill, 2002).
4. Adapted from Raiffa et al., 2002.
5. Prolonged negotiations in each individual act of piracy may be considered 'mutually hurting' due to factors discussed subsequently. Piracy as a whole is not hurting the pirates and the situation is therefore not 'ripe for resolution'. It has also been opined that the situation is also not hurting the governments and ship-owners. See for example Bateman, Sam, *Sea Piracy: Some Inconvenient Truths*, 2010.
6. McArdle, Megan, "Kidnapped by Pirates at Sea? Here's How Economics Can Save You," *The Atlantic*, 2012 accessed at http://www.theatlantic.com/business/archive/2012/05/kidnapped-by-pirates-at-sea-heres-how-economics-can-save-you/256828/ on 12 Jan 2013.
7. Ibid.
8. http://news.bbc.co.uk/2/hi/africa/7752813.stm accessed 15 Feb 2012.
9. De Groot, Olaf J., Rablen, M.D., Shortland, A., "Barrgh-gaining with Somali Pirates", *Economics of Security Working Paper 74*, Berlin: Economics of Security (2012).
10. Ibid. In the study the GDP of the country for each crew member is used as a proxy for determining the reference value for the crew. "At the extremes, the lowest value is found for the ten crew members on board the Kenyan-owned MV Miltzow captured in 2005 and the highest is found for the 30 crew members of the French-owned MV Le Ponant".
11. The study by EUSECON used as the basic reference for this section states that if the pirates are assumed to move last, the offer should equal V_I, so $r\varphi = V_I$. It is asserted by the authors that "Our results are qualitatively unaffected by who makes the final offer, but quantitatively, there is a benefit from being the last mover". This view is debatable. Though mathematically accurate, the statement probably does not take

into account the lives of the crew held as hostages. If the offer equalled what could be claimed from insurance, and simultaneously resulted in release of the hostages, the owners would accept it without any hesitation. Such a solution, calculated purely on the basis of the ship's value balanced with the amount due from insurance may still be realistic in case the hostages have died due to whatever circumstances during the period of negotiation.

12. In this case it is assumed that the value of the ship has reached its lowest value due to the depreciation and only the hostages are in effect being bargained for. This assumption is an oversimplification not to be taken literally.
13. This is unlikely in a scenario where there are multiple pirate gangs and different ship-owners. It is therefore strictly relevant only in case of a repeated game between the same set of players. The effect of killing one set of crew on the negotiation by other owners cannot however be discounted.
14. Violent reaction to an unsatisfactory offer may result from an urge to take 'revenge' on the hostages. This would be in line with the well known frustration aggression hypothesis. This however has a low probability in a situation under consideration and may be more applicable to a terrorist hostage situation/kidnapping.
15. This will also be intuitively apparent by asking the question "What would the pirates gain by killing the hostages?"
16. It is logical to assume that S will work on this premise. If this were not the case the owner would be willing to pay upto the maximum amount limited only by availability of funds.
17. This is a quadratic function of the form ($ax^2 + by + c$). If the same is taken as representing a concave quadratic function, the optimum value would be the inflection point on the curve.
18. Accordingly it may be seen that the aim of the policy makers must be to decrease the incentives to engage in acts of piracy. This may be achieved by a decrease in aor w and by an increase in q.
19. www.thenational.ae/news/uae-news/six-mv-orna-crew-still-held-by-pirates#ixzz2Lz HsqwPy accessed 10 February 2013.
20. Ibid.
21. IHS Jane's report, *Ship to shoreline: EU NAVFOR expands its anti-piracy mission*, (2012).
22. http://www.huffingtonpost.com/2012/09/01/somali-pirates-hostage-killed_n_1848937. html accessed 13 February 2013.
23. www.thenational.ae/news/uae-news/six-mv-orna-crew-still-held-by-pirates#ixzz2Lz L3bXpC accessed 10 February 2013.
24. http://www.nbcnews.com/id/41715530/ns/world_news-africa/#.USxbOX1xiRs accessed 13 February 2013.
25. http://www.thehindubusinessline.com/industry-and-economy/somali-pirates-continue-to-keep-indian-hostages-despite-ransom/article1701331.ece accessed 13 February 2013.
26. De Groot, Olaf J., Rablen, M.D., Shortland, A. "Barrgh-gaining with Somali Pirates," *Economics of Security Working Paper 74*, Berlin: Economics of Security (2012).
27. Ibid.

Bibliography and Further Reading

Books

Axelrod, Robert M., *The Evolution of Cooperation* (New York: Basic Books, 1984).

Binmore, K.G., *Game Theory: A Very Short Introduction* (New York: Oxford University Press, 2007).

Bunn, M.E., *Can Deterrence Be Tailored?* (Washington, D.C.: Institute for National Strategic Studies, National Defence University, 2007).

Burnett, John S., *Dangerous Waters: Modern Piracy and Terror on the High Seas* (New York, N.Y.: Dutton, 2002).

Casti, J.L., *Five Golden Rules: Great Theories of 20th-Century Mathematics and Why They Matter* (New York: Wiley & Sons, 1996).

Cohen, Steven, *Negotiating Skills for Managers* (New York: McGraw-Hill, 2002).

Davis, Paul K. and Jenkins, Brian M., *Deterrence and Influence in Counterterrorism* (Santa Monica, CA: Rand Corporation, 2002).

De Groot, Olaf J., Rablen, M.D., Shortland, A., *Barrgh-gaining with Somali Pirates* (Economics of Security Working Paper 74, Berlin: Economics of Security. 2012).

Enders, Walter, and Todd Sandler, *The Political Economy of Terrorism* (England: Cambridge University Press, 2006).

Freedom to Use the Seas: India's Maritime Military Strategy, Integrated Headquarters Ministry of Defence (Navy), 2007.

Gareau, Frederick H., *State Terrorism and the United States: From Counterinsurgency to the War on Terrorism* (Atlanta, GA: Clarity Press, 2004).

Goertz, Gary and Starr, Harvey, *Necessary Conditions: Theory, Methodology, and Applications* (Lanham, Maryland: Rowan & Littlefield Publishers, 2003).

Greenberg, Michael D., Chalk, Peter, *et al.*, *Maritime Terrorism: Risk and Liability* (Santa Monica, CA: RAND Centre for Terrorism Risk Management Policy, 2006).

Hoffman, Bruce, *Inside Terrorism* (New York: Columbia University Press, 2006).

Isaacs, Rufus, *Differential Games: A Mathematical Theory with Applications to Warfare and Pursuit, Control and Optimization* (New York: Wiley, 1965).

Kahn, Herman, *On Escalation: Metaphors and Scenarios*, Hudson Institute Series on National Security and International Order, No. 1 (New York: Praeger, 1965).

Myerson, Roger B., *Force and Restraint in Strategic Deterrence: A Game-Theorist's Perspective* (Carlisle Barracks, PA: Strategic Studies Institute, U.S. Army War College, 2007).

Myerson, Roger B., *Game Theory: Analysis of Conflict* (Cambridge, Mass: Harvard University Press, 1991).

Nisgav, Yair, *Some Game Theory Models for Allocating Forces in a Narrow Strait against Hostile Activity* (Monterey, Calif: Naval Postgraduate School, 1973).

Osborne, Martin J., and Ariel Rubinstein, *A Course in Game Theory* (Cambridge, Mass: MIT Press, 1994).

Osborne, Martin J., *An Introduction to Game Theory* (New York: Oxford University Press, 2004).

Payne, John D., *Fighting for Control: State-sponsored Terrorism as Foreign Policy in Cuba and Libya, 1959-2010, 2011* (Massachusetts Institute of Technology, 2011).

Payne, Keith B., *Deterrence in the Second Nuclear Age* (Lexington: University Press of Kentucky, 1996).

Pritchett, Clark W., *A Game Theory Approach to Search* (Ft. Belvoir: Defence Technical Information Centre, 1982).

Roberts, Brad, *Deterrence and WMD Terrorism: Calibrating Its Potential Contributions to Risk Reduction* (Ft. Belvoir: Defence Technical Information Centre, 2007).

Romp, Graham, *Game Theory: Introduction and Applications* (England: Oxford University Press, 1997).

Sakhuja, Vijay, *Asymmetric Warfare and Low Intensity Maritime Operations: Challenges for Indian Navy* (New Delhi: Observer Research Foundation, 2006).

Scott-Clark, Cathy and Levy, Adrian, *The Siege: Three Days of Terror Inside the Taj* (Viking, 2013).

Singh, Jaswinder, *Countering Small Boat Terrorism in Territorial Sea* (Monterey, California: Naval Postgraduate School, 2010).

Thomas C. Schelling, *Strategy of Conflict* (New York: Oxford University Press, 1963).

Thomas C. Schelling, *The Diplomacy of Violence* (New Haven: Yale University Press, 1966).

Thomas, L.C., *Games, Theory and Applications* (Mineola, N.Y: Dover Publications, 2003).

Williams, J.D., *The Compleat Strategyst: Being a Primer on the Theory of Games of Strategy* (New York: McGraw-Hill, 1966).

Journal Articles

Atzili, Boaz and Pearlman, Wendy, "Triadic Deterrence: Coercing Strength, Beaten by Weakness," *Security Studies*, 21:2 (2012): 301-335.

Azam, Jean-Paul, "Suicide-bombing As Inter-Generational Investment," *Public Choice*, 122 (2005): 1-2.

Bicr, V.M., Olivcros, S., & Samuclson, L., "Choosing What to Protcct: Strategic Defensive Allocation against an Unknown Attacker," *Journal of Public Economic Theory*. 9.4 (2007): 563-587.

Brodie, Bernard, "The Anatomy of Deterrence," *World Politics* 11, No. 2 (January, 1959): 173-191.

Chlebik, Kevin, "Terrorism and Game Theory: From the Terrorists' Point of View," *Pepperdine Policy Review*, Vol. III (2010): 15.

Dasgupta, S., and S.P. Cohen. "Is India Ending Its Strategic Restraint Doctrine?" *Washington Quarterly*, 34.2 (2011): 163-177.

Farrow, S., "The Economics of Homeland Security Expenditures: Foundational Expected Cost-Effectiveness Approaches," *Peace Research Abstracts Journal*, 44.4 (2007): 14-26.

Fearon, James D., "Rationalist Explanations for War," *International Organization*, 49, No. 3 (Summer, 1995): 379-414.

Gaver, D.P., K.D. Glazebrook, and P.A. Jacobs. "Search for a Malevolent Needle in a Benign Haystack," *International Series in Operations Research and Management Science*, 128 (2009): 125-146.

Gkonis, Konstantinos G., Psaraftis, Harilaos N. and Ventikos, Nikolaos P., "Game Theory Contributions to Terrorism Analysis in Merchant Shipping: An Application to Port Security," School of Naval Architecture and Marine Engineering, National Technical University of Athens.

Gladwell, Malcolm, "How David Beats Goliath: When Underdogs Break the Rules," *The New Yorker*, 85, No. 13 (May 11, 2009): 40.

Harsanyi, John C., "An Equilibrium-Point Interpretation of Stable Sets and a Proposed Alternative Definition," *Management Science*, 20.11 (1974): 1472-1495.

Jacobson, D., & Kaplan, E.H., "Suicide bombings and targeted killings in (counter-) terror games," *Journal of Conflict Resolution*, 51(5), 2007: 772-792.

Jain, S., & Mukand, S.W., "The economics of high-visibility terrorism," *European Journal of Political Economy*, 20(2), 2004: 479-494.

Dickerson, John P., Mannes, Aaron and Subrahmanian V.S., "Dealing with Lashkar-e-Taiba: A Multi-Player Game", *Theoretic Perspective.*

Kilgour, D. Marc; Brams, Stephen J., "The Truel", *Mathematics Magazine* 70 (5): 315–326 (December 1997).

Lapan, Harvey E., and Todd Sandler, "To Bargain or Not to Bargain: That is the Question," *The American Economic Review*, 78.2 (1988): 16-21.

Marlow, D.O., P. Kilby and G.N. Mercer, "The Travelling Salesman Problem in Maritime Surveillance – Techniques, Algorithms and Analysis."

Powell, R. "Defending against Terrorist Attacks with Limited Resources," *American Political Science Review*, 101.3 (2007): 527-542.

Pyeatt, Menefee S., "Piracy, Terrorism, and the Insurgent Passenger: A Historical and Legal Perspective," *Maritime Terrorism and International Law* (1990): 43-68.

Quackenbush, Stephen L., "General Deterrence and International Conflict:

Testing Perfect Deterrence Theory," *International Interactions*, 36, 2010: 60–85.

Rosendorff, B.P., and Todd Sandler, "Too Much of a Good Thing?: The Proactive Response Dilemma," *The Journal of Conflict Resolution: a Quarterly for Research Related to War and Peace.* 48.5 (2004): 657-671.

Sandler, Todd, and Daniel G.A.M., "Terrorism & Game Theory," *Simulation & Gaming*, 34.3 (2003): 319-337.

Sandler, Todd, and Harvey E. Lapan, "The Calculus of Dissent: an Analysis of Terrorists' Choice of Targets," *Synthese*, 76.2 (1988): 245-261.

Sevillano, Juan C., Insua D. Rios, and Jesus Rios. "Adversarial Risk Analysis: the Somali Pirates Case," *Decision Analysis*, 9.2 (2012): 86-95.

Siqueira, K., & Sandler, T., "Terrorist Backlash, Terrorism Mitigation, and Policy Delegation," *Journal of Public Economics,* 91(9), 2007: 1800-1815.

Stewart, Ian, "A Puzzle for Pirates," *Scientific American*, 280.5 (1999): 98-99.

Wintrobe, R., "Extremism, Suicide Terror, and Authoritarianism," *Public Choice*, 128 (2006): 169-195.

Zhuang, J., & Bier, V.M., "Balancing Terrorism and Natural Disasters—Defensive Strategy with Endogenous Attacker Effort," *Operations Research*, 55(5), 2007: 976-991.

Index